THE PARADOX OF PREDICTIVISM

An enduring question in the philosophy of science is the question of whether a scientific theory deserves more credit for its successful predictions than it does for accommodating data that was already known when the theory was developed. In *The Paradox of Predictivism*, Eric Barnes argues that the successful prediction of evidence testifies to the general credibility of the predictor in a way that evidence does not when the evidence is used in the process of endorsing the theory. He illustrates his argument with an important episode from nineteenth-century chemistry, Mendeleev's Periodic Law and its successful predictions of the existence of various elements. The consequences of this account of predictivism for the realist/anti-realist debate are considerable, and strengthen the status of the 'no miracle' argument for scientific realism. Another significant consequence is that scientific method embodies a pervasive epistemic pluralism, according to which expert scientists who assess theories depend heavily on the judgments of other scientists. Barnes's important and original contribution to the debate will interest a wide range of readers in philosophy of science.

ERIC CHRISTIAN BARNES is Associate Professor of Philosophy at Southern Methodist University.

THE PARADOX OF PREDICTIVISM

ERIC CHRISTIAN BARNES

Southern Methodist University

CAMBRIDGE
UNIVERSITY PRESS

CAMBRIDGE
UNIVERSITY PRESS

University Printing House, Cambridge CB2 8BS, United Kingdom

Cambridge University Press is part of the University of Cambridge.

It furthers the University's mission by disseminating knowledge in the pursuit of
education, learning and research at the highest international levels of excellence.

www.cambridge.org
Information on this title: www.cambridge.org/9780521879620

© Eric Christian Barnes 2008

First published 2008
First paperback edition 2011

A catalogue record for this publication is available from the British Library

ISBN 978-0-521-87962-0 Hardback

THE PARADOX OF PREDICTIVISM

ERIC CHRISTIAN BARNES

Southern Methodist University

CAMBRIDGE
UNIVERSITY PRESS

CAMBRIDGE
UNIVERSITY PRESS

University Printing House, Cambridge CB2 8BS, United Kingdom

Cambridge University Press is part of the University of Cambridge.

It furthers the University's mission by disseminating knowledge in the pursuit of
education, learning and research at the highest international levels of excellence.

www.cambridge.org
Information on this title: www.cambridge.org/9780521879620

© Eric Christian Barnes 2008

First published 2008
First paperback edition 2011

A catalogue record for this publication is available from the British Library

ISBN 978-0-521-87962-0 Hardback

To Veronica, who is my happiness

Contents

Acknowledgments

In the course of writing this book I have been helped by many people. My most immediately obvious debt of gratitude is to my colleagues at Southern Methodist University who have supplied me with considerable intellectual support and expertise during my thirteen years at SMU – most of which has been occupied with something or other to do with this book. These people include Roberta Ballarin, Phillipe Chuard, Alan Hausman, David Hausman, Mark Heller, Steve Hiltz, Robert Howell, Jean Kazez, Jim Lamb, Clayton Littlejohn, Mark McCullough, Alastair Norcross, Luke Robinson, Jonathon Sutton, Brad Thompson, and Steve Sverdlik. I am conscious of a special debt to Doug Ehring, who has been a terrific mentor during my time at SMU. My thoughts as presented in this book have also been the result of correction and/or nourishment by comments (in and out of print) by Peter Achinstein, David Christensen, Ellery Eells, Keith Lehrer, Patrick Maher, Paul Meehl, Elliot Sober, Edrie Sobstyl, Eric Scerri, Kent Staley, Wayne Woodward, and John Worrall. At Cambridge University Press I would like to thank Hilary Gaskin and two anonymous referees for helpful suggestions and comments on this work.

I am grateful to the National Endowment for the Humanities for a 1994 summer seminar stipend that kicked off my interest in confirmation theory, and to Larry Laudan for teaching an excellent seminar. Thanks are also due to Southern Methodist University for a sabbatical leave during the spring of 2005 during which part of this book was written.

This book draws on some of my published articles. These include the articles listed below. In every case I am indebted to journal referees who offered considerable help in working through my ideas. I am grateful to Oxford University Press, Springer Publishing Company, and Taylor and Francis Group for permission to include material from the following articles:

1996 "Discussion: thoughts on Maher's Predictivism," *Philosophy of Science*, 63, 401–410.

1996 "Social predictivism," *Erkenntnis* 45, 69–89.

1998 "Probabilities and epistemic pluralism," *British Journal for the Philosophy of Science* 49, 31–37.

1999 "The quantitative problem of old evidence," *British Journal for the Philosophy of Science* 50, 249–264.

2000 "Ockham's razor and the anti-superfluity principle," *Erkenntnis* 53, 3, 353–374.

2002 "Neither truth nor empirical adequacy explain novel success," *Australasian Journal of Philosophy* 80, 4, 418–431.

2002 "The miraculous choice argument for realism," *Philosophical Studies* 111, 2, 97–120.

2005 "Predictivism for pluralists," *British Journal for the Philosophy of Science*, 56, 421–450.

I would also like to take this opportunity to offer thanks to those who taught me philosophy. To begin at the all important beginning: Susan Mills Finsen, Daniel Graham, Dan Magurshak, and Jack Worley were certainly the four most important things that happened to me during my undergraduate years at Grinnell College. Their teaching, encouragement and support not only prepared me well and got me excited about philosophy, but helped me believe in myself at a time when such help was truly appreciated. The story continues at Indiana University, where I was likewise treated to wonderful teaching from many professors, but especially Nino Cochiarella, Alberto Coffa, Noretta Koertge, and John Winnie. I learned a lot from my colleagues at Denison University during my four years there. To all these people I am grateful for helping me grow as a philosopher, and for many fine memories of youth.

Reaching further back in time, my debts grow more profound even as they become harder to describe. My mother and father created a very loving and happy home in which the life of the mind was greatly admired. I suspect my fascination with scientific method has its roots in my memories of my father's laboratory, where my earliest memories of the world of grown ups were formed. At a tender age I came to connect (oversimply, obviously) the capacity for scientific thought with the living of a fully realized human life. I like to think that my attempt to understand scientific method is part of a larger project of understanding what the living of such a life entails.

I owe an immeasurable debt of gratitude to my wife, Veronica Barnes, whose love, kindness and wise counsel continue to dazzle me after nineteen years of marriage. And to my daughters, Rachel and Faith, I express heartfelt gratitude for sometimes quieting down when I needed to write, and for all the happiness and fun they bring to my life.

CHAPTER I

The paradox of predictivism

1.1 INTRODUCTION

Suppose that after years of living in genteel poverty, you have inherited a small fortune. Having little financial expertise yourself, you decide you are in need of a financial advisor who will help you invest your money wisely. You consult with two candidates, each of whom endorses a particular investment strategy. Each candidate's strategy is based on an account of the forces that induce the value of various investments to fluctuate. In fact, the two advisors can offer detailed explanations of why the value of these investments have changed in the way they have over the past five years. There is, however, one difference between the two advisors: one offered his account *prior* to the beginning of the five-year period, thus successfully *predicting* the various price changes. The other offered her account *after* the five-year period, and thus proposed to explain the price changes after they occurred. Now the question is whether you have, based on just this information, any reason to prefer one advisor over the other. One might insist that the two advisors are on equal ground: both offer accounts that are consistent with the same body of data. But it seems obvious – to many – that there is reason to prefer the advisor who made successful predictions over the one who didn't. If you agree, you may be inclined to endorse a particular view about how evidence confirms theory, a view known as 'predictivism.'

In philosophical parlance, predictivism asserts that, when E is evidence for T, E supports T more strongly when it is a novel confirmation of T than when it is not. Much ink has been spilled over the nature of novelty – but three primary accounts have been suggested: the temporal account, which claims that E is a novel confirmation of T when E is not known at the time T was proposed (Lakatos 1970), the heuristic account (Zahar 1973) which claims that E is a novel confirmation of T when E is not built to fit T, and the theoretical account (Musgrave 1974) which claims E is a novel confirmation for T when E is not explained by any theory other than T. The temporal account was rejected long ago because it made predictivism into a thesis that struck most philosophers as obviously false – for it is absurd to suggest that the confirming power of E literally depends upon the time at which E was discovered. The theoretical account should be dismissed, in my view, because it has the opposite problem: it renders predictivism into a trivial truth, for it is simply obvious that E will confirm T more strongly if there are no plausible competing explanations for T. This leaves the heuristic account of novelty which renders the predictivist thesis into a thesis that is both plausible but non-obvious – and this account will be assumed in the remainder of this chapter (I propose a new account of novelty in Chapter 2). Predictivism now proclaims that, where E is evidence for T, E confirms T more strongly when T was not built to fit E. Applied to our example, this asserts that the price fluctuations confirm the predictor's account of market forces more strongly than the non-predictors because we believe (or at least suspect) that the non-predictor built her account to fit the data, while the predictor didn't. Is this version of predictivism true? In this chapter I will survey some of the particularly prominent aspects of the long and tangled philosophical literature on this question. I begin with a few examples for the sake of motivation.

1.2 PUMPING UP THE INTUITION THAT PREDICTION MATTERS

Let us stipulate – for the moment – that when T is built to fit E, T 'accommodates' E – when T entails E but was not built to fit E, then T 'predicts' E.[1] The examples I cite fall into three categories, which I call dubious accommodations (in which the fact that T was built to fit E renders T 'fishy' in some clear sense), glorious predictions (in which the fact that T predicts E seems extremely impressive evidence for T), and two popular thought experiments.

1.2.1 *Dubious accommodations (also known as ad hoc theory rescues)*

French bread

Let hypothesis H be: All bread is nourishing. H is believed to be supported by many cases in which it is known that H has proven true. However, it

---

[1] I am obviously proposing to use the locution 'T predicts E' in a technical sense that does not correspond to the common meaning of this expression which simply means 'T entails E.'

turns out that bread produced in a particular region in France is not nourishing – it is in fact poisonous. Thus H is modified to H': All bread – except that grown in the relevant region, which is poisonous – is nourishing. The modification of H into H' is an attempt to 'rescue' H from falsification, and H' is of course consistent with all known data. Nonetheless, despite this consistency, there seems to be something fishy about H'. The precise nature of the fishiness is, of course, the point in question here. But there is at least a well-established name for the fishy quality – the modification of H into H' is deemed an 'ad hoc theory adjustment.' It is a straightforward example of a dubious accommodation.

The levity of phlogiston

The chemical revolution of the eighteenth century involved the replace-ment of the phlogiston chemistry, supported by scientists like Priestley and Cavendish, with the "new chemistry" of Lavoisier and others. Proponents of the phlogiston theory had argued that phlogiston was a common component of many substances that was emitted in combustion. The phlogiston theory faced an anomalous result, however, in that some sub-stances actually gained weight when burned. Some proponents of the phlogiston theory, like Gren, responded by suggesting that phlogiston had negative weight – thus the process of emitting phlogiston could explain a corresponding weight gain. This attempt to rescue the phlogiston theory from refutation seemed to many scientists at the time (even to some supporters of the phlogiston theory) to be most unconvincing, despite its ability to reconcile the theory to the data (for discussion see Kitcher 1993: 272–290). Here again we encounter the fishy quality – this time in the form of a hypothesis conjoined with the phlogiston theory to protect that theory from falsification. It is referred to as an 'ad hoc hypothesis' – another form of dubious accommodation.

1.2.2 Glorious successful predictions

Psychic

You encounter a self-identified psychic who claims to have knowledge of your personal future. Despite no prior acquaintance with you, she makes a long and precise set of predictions P about your future. P is subsequently fully confirmed – to your utter astonishment. You conclude that the bizarre hypothesis of her clairvoyance is reasonably confirmed. The confirmation of P is an example of what I call a glorious successful prediction – a prediction so bold as to constitute, once confirmed, strong evidence for the hypothesis that predicted it.

Retrograde motion

Nicholas Copernicus proposed a heliocentric theory of the solar system in his 1543 work *De Revolutionibus Orbium Coelestium*. This theory provided a straightforward explanation of the phenomenon of retrograde motion (in which the planets periodically reverse their apparent motion against the fixed stars). By contrast, Ptolemaic astronomy had managed to accommodate retrograde motion only by positing a complex system of epicycles that were built to fit the data. The fact that Copernican astronomy required no elaborate 'fixing up' to accommodate retrograde motion has been equated with the claim that Copernican astronomy predicted retrograde motion (e.g., Scerri and Worrall 2001: 423), and therefore was, by all appearances, more strongly confirmed by it.

1.2.3 Two thought experiments

In each of the following two thought experiments there is a theory T and some evidence that supports T – in one scenario T is built to fit the evidence and in the other it isn't. The question is whether the degree of confirmation offered by E to T is the same in the two scenarios.

Connect the dots

The following example is discussed in Howson (1988: 381–382): A coordinate graph represents the relationship between variables x and y – in the first scenario, a set of data points E is plotted on the graph, and the smoothest possible curve is subsequently drawn to fit E – this curve is T. T is a bizarrely irregular curve. In the second scenario, the same curve T is drawn before the accumulation of data – when data is accumulated, E is established, and E falls neatly on T. The question is in which of the two scenarios, if either, is T better confirmed?

Coin flip

This example, due to Maher (1988), also contains two scenarios: in the first scenario, a subject (the accommodator) is presented with E, which is the outcome of a sequence of 99 flips of a single fair coin. E forms an apparently random sequence of heads and tails. The accommodator is then instructed to tell us the outcome of the first 100 flips – he responds by reciting E and then adding the prediction that the 100th toss will be heads – the conjunction of E and this prediction about the last toss is T. In the other scenario, the subject (the predictor) is asked to predict the first 100 flip outcomes without witnessing any outcomes – the predictor endorses theory T. Thereafter the coin is flipped 99 times, E is established, and the predictor's first 99 predictions are confirmed. The question is in which of these two scenarios is T better confirmed.

These stories all point to the plausibility of predictivism. But before we declare the issue settled, it should be noted there are other examples in which accommodated evidence can provide compelling evidence for a theory – I will call these 'glorious accommodations.' To borrow an example from Mayo (1996: 271), let E contain a list of the SAT scores of all the students in a particular class – a simple averaging technique can be used to generate (T). The average SAT score is x (for some particular value of x). Now clearly T was built to fit E – but E provides compelling evidence for T. Accommodation can, in some cases, provide perfectly compelling evidence for a theory (cf. Hobbes (1993)). Thus the predictivist thesis does not seem to be universally true. Things are turning out to be a bit complex.

1.3 THE PARADOX OF PREDICTIVISM

Predictivism is essentially a comparative thesis – it compares the probative weight of E as accommodated with the probative weight of E as predicted. This is why actual historical examples of dubious accommodations and glorious predictions provide only limited evidence for the truth of predictivism, for any actual historical case presents E in only one of the two roles. We are forced to ask, in the case of the levity of phlogiston for example, how well phlogiston theory would have been confirmed if the increased weight of certain burning substances had somehow been predicted by the phlogiston theory. This forces us to construct in the imagination a counterfactual scenario in which this prediction holds – but it is far from clear how we are to imagine the modified phlogiston theory (for there are a variety of ways in which phlogiston theory could be imagined to entail it). Hence the attractive quality of the two thought experiments, in which it seems easy to keep the two scenarios the same except for changing the role of E.

It is high time to note that as an account of how theories are evaluated in science, predictivism is deeply controversial. This is primarily because predictivism makes facts about how theories were constructed relevant to the epistemic assessment of theories. Specifically, predictivism entails that it matters with respect to the assessed probability of T whether it was built to fit any of its supporting evidence. But this makes the assessed probability of T curiously dependent on the mental life of its constructor, specifically on the knowledge and intentions of that theorist to build a theory that accommodated certain data rather than others. This means that scientists who want to assess the probability of some theory need to do more than simply know all the relevant evidence and criteria of theory assessment and apply them to the theory – they need to know facts about the biography of the theorist who constructed the theory. In what follows I will use the term 'biographicalism' to denote the view that facts about the life stories of scientists are epistemically relevant to the assessment of theories. Predictivism as defined here entails the truth of biographicalism, and this is why many philosophers reject it.

Philosophical literature on this subject abounds with passionate rejections of biographicalism. For one thing, if it were the case that biographical information were crucial for the proper evaluation of theories, then such proper evaluation would be difficult if not impossible for many working scientists to perform, given that the relevant biographical information (about what the scientist who constructed a theory knew and intended) is not always provided in the scientific literature that such scientists read, and can be unearthed only by subsequent historians of science who study the archives of theorists. Clearly "the fruits of such historical research are usually unavailable to the scientists involved in an ongoing rivalry between research programmes. The decisions of most of them about which programme to work on or believe in cannot, of course, be influenced by discoveries made by historians examining unpublished (sometimes oral) information – many years, perhaps after the rivalry has ended" (Gardner 1982: 6). Furthermore biographical facts seem to be widely ignored in scientific literature, except where issues of priority are concerned (Thomason 1992: 195). But even more to the point, the very idea of biographicalism flies in the face of what appear to be obvious facts about the objectivity of scientific method. "The extent to which a given postulate is confirmed by the evidence at hand is a function of the precise contents of that postulate and the nature of that evidence, and definitely not of such fortuitous factors such as the time when someone happened to hit upon the idea of formulating that postulate, or the stage at which it has in practice been employed and for what purpose" (Schlesinger 1987: 33). Leplin similarly points to the absurdity of biographicalism: "The theorist's hopes, expectations, knowledge, intentions, or whatever, do not seem to relate to the epistemic standing of his theory in a way that can sustain a pivotal role for them ..." (1997: 54). (See also Collins (1994) and Achinstein (2001: 210–230).)

A paradox is a statement that appears as though it cannot be true and yet somehow seems as though it must be true. Predictivism is genuinely paradoxical: it seems too obvious to question when we consider examples like those adduced in the previous section, but it seems too absurd to even consider when we consider its commitment to biographicalism as an account of scientific method. In the next section we consider some of the key moments in the history of the vast philosophical literature on predictivism with the hope that the paradox may be illuminated.

1.4 A SKETCHY HISTORY OF PREDICTIVISM

1.4.1 John Herschel, William Whewell, and the method of hypothesis[2]

There was in the eighteenth century a passionate debate about scientific method. Newton's *Principia* (1687) had famously repudiated the use of hypotheses in science, and many subsequently followed Newton in holding that the only theories worth considering were those that did not postulate unobservable entities but were straightforwardly 'deduced from the phenomena.' But other scientists defended the method of hypothesis by which theories about unobservables were postulated and shown to save the phenomena. One important proponent of the method of hypothesis was David Hartley, whose *Observations of Man* (which first appeared in (1749)) claimed that the brain and nervous system are filled with a 'subtle fluid' – or aether – which transmits vibrations from one point in the perceptual system to another. Hartley used this basic picture to concoct explanations of a vast array of phenomena, including sleep, the generation of simple and complex ideas, paralysis, taste, sexual desire, memory, and many more. Hartley defended his theory by pointing to the great number of phenomena which it could explain. He writes:

Let us suppose the existence of the aether, with these its properties, to be destitute of all direct evidence, still, *if it serves to explain a great variety of phenomena, it will have an indirect evidence in its favour by this means.* (*Observations of Man*, London, 1791, vol. I, p. 15, quoted in Laudan 1981: 115, italics in Laudan)

Hartley defends the method of hypothesis using a number of arguments, including appeals to its heuristic value (it can lead to the discovery of new phenomena and generate new experiments) and the fact that the method

---

[2] For my account in this section I have relied heavily on Laudan (1981: 111–140).

can produce hypotheses which, though not fully proven, are nonetheless 'the best science can do' in certain contexts.

Hartley's defense of the method of hypothesis met with scathing criticism. For as his critics pointed out "there were many rival systems of natural philosophy which – after suitable *ad hoc* modifications – could be reconciled with all the known phenomena. The physics of Descartes, the physiology of Galen, and the astronomy of Ptolemy would all satisfy Hartley's criterion. There was, in Hartley's approach to the epistemology of science, nothing which would discredit the strategy of saving a discarded hypothesis by cosmetic surgery or artificial adjustments to it" (Laudan 1981: 117). This had, after all, been the point of Newton's repudiation of the method of hypothesis.

But something fundamental happened in the early nineteenth century to change all this. The story involves a glorious prediction. At this time most physicists held the 'emission theory' of light which claimed that light consisted of particles. However in 1819 the young French physicist Augustin Fresnel wrote an essay propounding a version of the wave theory of light and showed that various known facts supported his theory. He submitted his essay to a science competition held by the French Academy. One of the prize commission's judges, Poisson, demonstrated that Fresnel's wave theory of light had an obviously absurd consequence: it implied that if a small circular disk were used to create a shadow cast by light from a small hole, the center of the disk's shadow would contain a spot of bright light. This consequence was tested by Arago, who with much amazement found the bright spot to be exactly where Fresnel's theory entailed it should be. Fresnel won the prize in question. According to a common version of the story, this glorious successful prediction proved a turning point in the physics community with respect to the nature of light, and by the mid to late 1820s the wave theory was generally accepted in the scientific community.[3]

Light waves, of course, are unobservable entities – and thus the wave theory of light had been formulated by an application of the method of hypothesis. But in this case, unlike that of Hartley's aether theory of the nervous system, the theory had led to an astonishing successful prediction. This led individuals like John Herschel and William Whewell to argue that the method of hypothesis was legitimate when it produced hypotheses

---

[3] For discussion of relevant points, including the shortcomings of this 'general version' of the story, see Worrall (1989). See also Achinstein (1991 Part 1), which argues that the methods used by proponents and opponents of the Method of Hypothesis actually had many commonalities.

that led to the successful prediction of previously unknown phenomena. Successful prediction thus served to distinguish the arbitrary and ad hoc products of hypothesizing like Hartley's aetherial account of the nervous system from legitimate products like the wave theory of light. In this context Whewell and Herschel embraced predictivism. Insofar as the wave theory was widely accepted by the scientific community, it seemed as though Whewell's and Herschel's predictivism was also widely accepted.

There was just one problem with the status of predictivism in the early nineteenth century: according to Laudan, neither Whewell nor Herschel offered any logical or epistemological explanation of why it held true (1981: 131, 134). There was no attempt on their part, e.g., to explain the oddly biographical picture of scientific method it entailed. It was simply offered as a rule that scientists should follow.[4] Whewell and Herschel thus advocated predictivism while ignoring the obvious problems it raised. John Stuart Mill in his *System of Logic* responded to Whewell by emphatically denying that any special significance should be assigned to the fact that an established consequence of a theory was predicted rather than accommodated (1961: 328–329). The paradox of predictivism had emerged – but no clues about its resolution were yet to be found.

1.4.2 Popper, Lakatos, and the fallacy of pure consequentialism

Karl Popper is probably the most famous proponent of the preference for prediction in the entire history of philosophy. In his lecture "Science: Conjectures and Refutations" Popper recounts his boyhood attempt to grapple with the question '*When should a theory be ranked as scientific?*' (Popper 1963: 33–65). Popper had become convinced that certain popular theories of his day, including Marx's theory of history and Freudian psychoanalysis, were pseudosciences. Though they superficially resembled genuinely scientific theories, like Einstein's theory of relativity, they fell short of scientific status. Popper deemed the problem of distinguishing scientific from pseudoscientific theories the 'demarcation problem.' His solution to the demarcation problem, as everyone knows, was to identify the quality of falsifiability (or 'testability') as the mark of the scientific theory.

---

[4] Worrall (1985: 324) responds to Laudan by claiming that Whewell did purport to give a rationalization of predictivism along the following lines: predictions carry special weight because a theory that correctly predicts a surprising result cannot have done so by chance, and thus must be true. Insofar as this counts as an argument at all it appears to be a version of what I call the miracle argument for strong predictivism – I discuss this argument and argue that it is fallacious in Chapter 4.

The pseudosciences were marked, Popper claimed, by their vast explanatory power. They could explain not only all the relevant actual phenomena the world presented, they could explain any *conceivable* phenomena that fell within their domain. This was because the explanations offered by the pseudosciences were sufficiently malleable that they could always be adjusted ex post facto to explain anything. Thus the pseudosciences never ran the risk of being inconsistent with the data. By contrast, a genuinely scientific theory – such as Einstein's theory of relativity – made specific predictions about what should be observed and thus ran the risk of falsification. Popper emphasized that what established the scientific character of relativity theory was that it 'stuck its neck out' in a way that pseudosciences never did.

At one level all of this should strike us as familiar. The theories Popper identifies as pseudoscientific are strikingly reminiscent of Hartley's aetherial theory discussed above, which likewise made no specific predictions but simply offered ex post facto explanations of phenomena it was built to fit. Like Whewell and Herschel, Popper appeals to the predictions a theory makes as a way of separating the illegitimate uses of the method of hypothesis from its legitimate uses. But there was a big difference as well. Whewell and Herschel pointed to predictive success as a necessary condition for the acceptability of a theory that had been generated by the method of hypothesis. Popper by contrast focuses in his solution to the demarcation problem not on the success of a prediction but on the fact that the theory made the prediction at all – as noted above, what marked a theory as scientific was not that its detailed predictions were confirmed, but simply that it made specific predictions which established a risk of falsification. This criterion of scientific status would have worked to disqualify Hartley's aetherial theory as well as it did to disqualify Freudian psychoanalysis.[5] Of course, there was for Popper an important difference between scientific theories whose predictions were confirmed and those whose predictions were falsified. Falsified theories were to be rejected, whereas theories that survived such testing were to be 'tentatively accepted' until such time as they might be falsified. Popper did not hold, with Whewell and Herschel, that successful predictions could constitute legitimate proof of a theory – in fact, Popper held that it was impossible to show that a theory was even probable on the basis of the evidence, for he embraced Hume's critique of inductive logic which made evidential

---

[5] Popper's argument that Freudian psychoanalysis is unfalsifiable has been questioned by Grunbaum (1984).

support for the truth of theories impossible. Thus it is incorrect to ascribe to Popper a commitment to predictivism, as predictivism entails the possibility of inductive support. Despite this, he certainly sounded as though he was endorsing predictivism when he wrote that "Confirmations should count only if they are the result of *risky predictions*; that is to say, if, unenlightened by the theory in question, we should have expected an event which was incompatible with the theory – an event which would have refuted the theory." (1963: 36) It would ultimately prove impossible for Popper to reconcile his claim that a theory which enjoyed predictive success ought to be 'tentatively accepted' with his anti-inductivism (see, e.g., Salmon 1981).

What argument did Popper offer for the preference for prediction? Some have claimed that Popper offered no argument at all (e.g., Laudan 1981: 134), but I do not agree. Popper linked the preference for prediction to his claim that a theory is scientific to the extent that it forbids certain observable states of affairs. The more a theory forbids, the better it is. He then linked the ad hoc modification of a theory to its diminished falsifiability:

Some genuinely testable theories, when found to be false, are still upheld by their admirers – for example by introducing *ad hoc* some auxiliary assumption, or by re-interpreting the theory *ad hoc* in such a way that it escapes refutation. Such a procedure is always possible, but it rescues the theory from refutation only at the price of destroying, or at least lowering, its scientific status. (1963: 37)

If we equate the process of building a theory to fit known data with the sort of *ad hoc* moves Popper refers to here, and if it is the case that such ad hoc moves serve to generate theories with low falsifiability, and if it is the case that a theory is 'better' the more falsifiable it is, then it follows that there is something systematically disreputable about theories that are built to fit the facts. This certainly seems to be an argument for a preference for prediction, if not for predictivism per se.

However, Popper's argument falls apart on careful inspection. First of all, Popper's claim that the process of modifying a theory ad hoc has a systematic tendency to reduce its falsifiability is never clearly established. As an example Popper points to Marxist theory, which on his analysis began as a falsifiable theory but was modified by its proponents so as to become unfalsifiable. But we are given no reason to think that ad hoc modifications are in general prone to diminish falsifiability, nor is there any particular reason to think this. E.g., in the case of the example of *French bread* (1.2.1), the original hypothesis that all bread is nourishing does not forbid any more possible observable states of affairs than the modified

hypothesis that all bread nourishes except that grown in a particular region
of France, which is poisonous. Bamford makes essentially this point when
he objects that "Popper conflates the idea of a hypothesis that is formulated
only with the aim that it should entail some observation statement O with
the idea of a hypothesis that does not entail any observation statement
other than O (and its entailments), that is, a hypothesis that is not
independently testable of O)." (1993: 336) But these are clearly distinct
properties of a hypothesis and there is no reason to presume a general
correlation between them.[6] Bamford also argues persuasively that there is
no reason to think that the replacement of a testable theory with another
theory with diminished empirical content constitutes unscientific practice
in general (1993: 350), contrary to Popper's solution to the demarcation
problem. Finally, Popper's insistence that there is *absolutely no reason* to
think that theories that survive rigorous testing are true ultimately undercuts his own solution to the demarcation problem. For if there is literally
no reason to think that such theories are even a little bit probable, there
is no real basis for preferring such theories to pseudoscientific rivals from
the standpoint of their respective credibility. Einstein's theory of relativity,
no matter how many risky predictions it might successfully make, is no
more likely than the myths of Homer to represent the truth about, say, the
motion of heavenly bodies. Popper is left with no reason to sanction the
scientific community's ongoing engagement with one rather than the other.
Popper's attempt to ground a preference for prediction is multiply flawed.

Imre Lakatos (1970, 1971) proposed an account of scientific method
in the form of his well-known 'methodology of scientific research programmes' (MSRP). A scientific research programme was constituted by a
'hard core' of propositions which were retained throughout the life of
that programme together with a 'protective belt' which was constituted
by auxiliary hypotheses that were adjusted so as to reconcile the hard core
with the empirical data. The attempt on the part of the proponents of the
research programme to reconcile the programme to empirical data produces a series of theories $T_1, T_2, \ldots, T_N$ where, at least in some cases, T_{i+1}
serves to accommodate some data that is anomalous for T_i. Lakatos held
that a research programme was 'theoretically progressive' insofar as each
new theory predicts some novel, hitherto unexpected fact. A research

---

[6] An indication of how influential Popper's coupling of 'built to fit the data' and 'untestable' has been
is contained in the definition of 'ad hoc hypothesis' in the *Dictionary of the History of Science*:
"Hypotheses, designed for a specific purpose, that accomplish nothing else ... *ad hoc* hypotheses
typically have no new content open to any independent check; still less are they testable in genuinely
new ways." (Bynum et al., 1981: 6–7).

programme is 'empirically progressive' to the extent that its novel empirical content was corroborated, that is, if each new theory leads to the discovery of "some *new fact.*" (Lakatos 1970: 118) Lakatos thus offered a new solution to the demarcation problem: a research programme was pseudoscientific to the extent that it was not theoretically progressive. Theory evaluation is construed in terms of competing research programmes: a research programme defeats a rival programme by proving more empirically progressive over the long run.

Lakatos' MSRP thus followed Popper's falsificationist methodology in two crucial respects (among others): (1) risky predictions constitute the mark of scientific character (and, relatedly, the accommodation of data is scientifically suspect), and (2) successful predictions – aka novel confirmations – are the sole mark of scientific success. But these are not really points for which Lakatos argued – they are simply taken over from Popperianism as axioms on which the MSRP is built. The objections noted above against Popper's endorsement of (1) and (2) apply no less to Lakatos. Hacking writes that "Lakatos sides with Whewell against Mill, but he does not give reasons. Rather he makes it true by definition that what matters to a theory is its ability to predict novel facts. For that is what he comes to mean by 'progressive' . . ." (1979: 389). Lakatos says nothing that illuminates the paradox of predictivism.

Zahar, an early proponent of MSRP, recognized that it was not inevitably the case that an ad hoc hypothesis – one built to fit the data – would fail to be more testable than its predecessor. But there were, he thought, other ways in which the quality of ad hocness served to impugn a hypothesis. He distinguished three kinds of ad hoc-ness:

A theory is said to be *ad hoc$_1$* if it has no novel consequences as compared to its predecessor. It is *ad hoc$_2$* if none of its novel consequences have been actually 'verified'; . . . [A theory is] *ad hoc$_3$* if it is obtained from its predecessor through a modification of its auxiliary-hypotheses which does not accord with the spirit of the heuristic of the programme. (1973: 101)

Nickles (1987) offers several astute observations about these three criteria. The first is that there is a severe tension between the first two and the third – the first two follow the Popperian injunction toward greater empirical content and theory corroboration. The third, while it means to block the sort of arbitrary adjustment of theory to data that Popper denounced, actually forbids any heuristically unmotivated theory change. "Hence, blind guesses and Popperian conjectures turn out to be *ad hoc$_3$*, regardless of their novel predictive power!" (Nickles 1987: 192). There is no apparent rationale for disqualifying ad hoc$_3$ hypotheses from the standpoint of MSRP. But for that matter, there is no apparent rationale for disqualifying ad hoc$_2$ hypotheses on grounds of their novel consequences being unverified, for MSRP follows Popper's anti-inductivist position that consequence verification offers *literally zero* support. "Because MSRP retains a Popperian conception of empirical support, it fails to provide an adequate motivation for heuristics and is ultimately incoherent." (Nickles 1987: 181–182) As for ad hoc$_1$ hypotheses, we have noted above that there is no reason to assume that the replacement of a theory with a less testable successor is in general unscientific. We are left with no reason to look askance on any of the forms of adhoc-ness identified by Zahar. We are left, in short, with the paradox of predictivism unresolved.

Nickles (1988) offers an insightful diagnosis of the preference for prediction so popular among philosophers of science like Whewell, Popper, and Lakatos. The psychogenesis of this preference is attributable to a view that Nickles calls the 'Central Dogma' of confirmation theory, which is:

All empirical support (of a theoretical claim) = empirical evidence = empirical data = successful test results = successful predictions or postdictions = true empirical consequences (of the claim plus auxiliary assumptions). (1988: 393)

Nickles deems this view 'pure consequentialism' because it affirms that nothing except the truth-value of a theory's empirical consequences is of any relevance to an assessment of a theory's credibility. Any pure consequentialist, he notes, faces the methodological predicament posed by the underdetermination of theory by data: for any body of data D, there are multiple theories that fit D equally well. Appealing to non-consequential forms of theory support is the most reasonable solution to the underdetermination problem: forms of such support would include the application of extra-empirical criteria such as simplicity, elegance or fruitfulness, or coherence with background knowledge. Although multiple theories may all fit the data equally well, the application of such non-consequential criteria may serve to greatly delimit the competitors, in some cases down to a single theory. But such non-consequential forms of support are not available to pure consequentialists like Popper and Lakatos. Hence Popper and Lakatos turned to methodology as the way out of the methodological predicament: since one could not disqualify theories that fit the data but were nonetheless clearly disreputable in epistemic terms, one must disqualify them on grounds of the method by which they were built – they were constructed so as to fit the facts, and this was their putative shortcoming. Thus pure consequentialists are led to biographicalism: facts about the mental lives of theorists were embraced as relevant to theory assessment, for there was no other apparent way around the underdetermination of theory by data.

But pure consequentialism is an utterly wrongheaded picture of theory assessment (as Nickles argues with great plausibility).[7] On the assumption that science is currently possessed of any theoretical knowledge at all, and on the further assumption that scientists know (or at least reasonably believe) that certain theories are approximately correct, then any other theory that is deduced from or supported by such knowledge will qualify as either known or supported by a non-consequential form of evidence. For that matter, the evolution and application of various forms of extra-empirical criteria in theory assessment has long been a well-charted story in the history and philosophy of science (for one excellent example, see McAllister 1996). There is no need to fall back on biographicalism to solve the methodological predicament. It suffices to reject pure consequentialism and embrace the usual forms of non-consequential support. The problem with Popper's pseudosciences, in short, is not that they were built to fit the data, but that they had nothing to recommend them except the fact that they fit some body of data salient in the operative context. They enjoyed no (or too little) non-consequential support. Mere consequential support is powerless to win the conviction of the scientific community – for there is always the underdetermination problem to be reckoned with. The Popperian and Lakatosian preference for prediction, in sum, rested on nothing more substantial than the fallacy of pure consequentialism. We are left with the paradox of predictivism unresolved.

1.4.3 The Keynesian dissolution of the paradox

In a brief passage in his book *A Treatise on Probability* (originally published in 1921), John Maynard Keynes provides a straightforward resolution of the

---

[7] Nickles (1987, 1988) uses the term 'generative support' to refer to the form of support that is (or could be) used to generate or discover a new theory, and explains that generative support involves reasoning *to* a theory, while consequential support involves reasoning *from* a theory to its consequences. He argues at length that the proper evaluation of theories requires the combination of generative and consequential support. I find this distinction somewhat obscure, however, as it seems to ignore the fact that support could be both consequentialist and generative when such support involves showing a theory to be 'discoverable' from its known consequences. I thus use the term 'non-consequential support' to refer to the form of support that is the proper foil of consequential support – 'non-consequential support' refers to the use of extra-empirical criteria and coherence with background knowledge.

paradox of predictivism. Predictivism, he argues, is false despite its prima facie plausibility. He explains:

The peculiar virtue of prediction or predesignation is altogether imaginary . . . The plausibility of the argument [for predictivism] is derived from a different source. If a hypothesis is proposed *a priori*, this commonly means that there is some ground for it, arising out of our previous knowledge, *apart from* the purely inductive ground, and if such is the case the hypothesis is clearly stronger than one which reposes on inductive grounds only. But if it is a mere guess, the lucky fact of its preceding some or all of the cases which verify it adds nothing whatever to its value. It is the union of prior knowledge, with the inductive grounds which arise out of the immediate instances, that lends weight to any hypothesis, and not the occasion on which the hypothesis is first proposed. (1929: 305–306)

By 'the inductive ground' for a hypothesis Keynes clearly means the data that the hypothesis fits. Keynes means that when a hypothesis is first proposed by some theorist who undertakes to test it, some other (presumably non-consequential) form of support prompted the proposal. Thus hypotheses which are proposed without being built to fit the empirical facts which they entail are typically better supported than hypotheses which are proposed merely to fit the data – but this does not reveal any qualitative difference between predicted and accommodated evidence. Once this is appreciated, the paradox of predictivism is dissolved: the paradox of predictivism arose only because the role of the preliminary proposal-inducing evidence had been ignored. Once the role of the latter is acknowledged, we can see that predictivism is false.

Colin Howson has been a vigorous proponent of the Keynesian dissolution. In his (1988: 382) he uses the example I deemed *Connect the dots* (1.2.3) to illustrate the Keynesian dissolution: our intuition that line T is better confirmed when it is proposed prior to the demonstration of the data points that fall on T than when T is built to fit E arises because in the former case we presume that T was proposed originally on some basis other than E. That other basis, whatever it is, combines with evidence E to generate overall more support for T than is available in the scenario in which T is built to fit E. One way of couching the Keynesian dissolution in a Bayesian vocabulary is to say that, in a case like *Connect the dots*, theory T has a higher prior probability (given the total available evidence) in the prediction scenario than it has in the accommodation scenario – given that the probative weight of E is essentially the same in both scenarios, the posterior of T emerges in the prediction scenario with a higher value than it has in the accommodation scenario. Thus predictivism seems true, but only to people who cannot see that the real problem with T in the accommodation scenario is that it has a low prior. The same approach could apparently handle *Coin flip*.

Howson (1984) applies what amounts to the Keynesian approach to another example that Zahar (1973: 257) had used to defend the Lakatosian prediction preference; the argument is refined in Howson (1988: 389) and (1990: 236–237). This example concerned the ability of Einstein's general theory of relativity (GTR) together with widely accepted auxiliary hypotheses (A) to predict the Mercury perihelion (M), which was widely considered to be an important piece of evidence for GTR. Newtonian mechanics (NM), together with A, did not entail M, but could be made to fit M by replacing A with a suitable auxiliary hypothesis (A*) which postulated an otherwise unsupported irregular shape to the Sun. Zahar had argued that M more strongly corroborated GTR than NM because, in some obvious sense, NM was reconciled to M by an ad hoc rescue that GTR did not need. The gist of Howson's response is as follows: M offered no actual confirmation to NM itself – since NM and M were probabilistically independent.[8] At best M raised the probability of NM & A* to the prior probability of NM (it thus did provide a little support for NM & A*, Zahar to the contrary). But insofar as GTR entailed M (given that A was part of background knowledge), M did more strongly support GTR than NM. To put the point another way, assuming the prior probabilities of GTR and NM at the time in question were roughly equal, the effect of M was to produce a higher posterior probability for GTR & A than for NM & A* – but this simply reflects a higher prior for GTR & A. This higher prior resulted simply from the fact there was a greater body of preliminary evidence that supported A than supported A*. Once we have taken into account the differences in the effects of preliminary evidence, the illusion of predictivism is dispersed – as Keynes argued long ago.

To put the prior example in slightly different terms, the hypothesis 'A*' which the Newtonian needs to reconcile NM to M is clearly an ad hoc hypothesis. The Keynesian approach offers a straightforward explication of what made A* ad hoc – it was that, despite the fact that it entailed a true consequence when conjoined with NM, it was otherwise badly supported.

---

[8] I believe there is a small glitch in Howson's analysis here. The claim that NM and M are probabilistically independent seems false if the auxiliary hypothesis A (that, conjoined with GTR, entailed M) was part of background knowledge. For NM & A entailed ~M, which means that M served to disconfirm NM. However, Howson does not really need the assumption that NM and M are probabilistically independent to make his point – he needs only the claim that M did not confirm NM while it did serve to raise the posterior (i.e. M-based) probability of NM & A* to the same value as the posterior probability of NM, together with the claim (adopted in the passage below) that the prior (i.e. M-free based) probabilities of GTR and NM were roughly equal.

This notion of the essence of the ad hoc hypothesis resembles the proposals of Schaffner (1974: 67–73) and Leplin (1975).[9] The ad hoc hypothesis, in other words, is best described not simply in terms of how it was constructed but as a consequence-fitting hypothesis that is nonetheless badly supported. But the fact that hypotheses could have true empirical consequences while nonetheless being hopelessly unsupported was the moral of the previous section, in which the poverty of pure consequentialism was noted. But in the previous section the fallacy of pure consequentialism was noted only to reveal the confused motivations of the Popperian and Lakatosian preference for prediction. Now it seems that the fallacy of pure consequentialism is connected to a serious proposal about how to resolve the paradox of predictivism – one that I have deemed the Keynesian dissolution. It is, it seems to me, the first genuine step forward toward the resolution of this paradox we have encountered so far. Predictivism, from this standpoint, is simply false.

1.4.4 Worrall: conditional and unconditional confirmation

John Worrall has elaborated a theory of confirmation that purports to show what truth there is in predictivism.[10] He claims that a typical scientific theory can be analyzed into a conjunction of a 'core' – a basic idea that characterizes that theory – and additional auxiliary claims that specify the theory's more particular content. Worrall's position is that in many cases accommodated evidence offers only conditional confirmation for the theory: if there is independent evidence that supports the core, the accommodated evidence supports the theory. Absent such independent evidence, accommodated evidence provides little or no support for the theory. Predicted evidence, however, often supplies direct evidence for the core of the theory, and thus provides unconditional confirmation of the theory. Worrall illustrates how his account works by considering the attempt on the part of some creationists to reconcile their account of the earth's history with the fossil record. The nineteenth century English creationist Phillip Gosse proposed that God created the fossil record to test our faith. The core idea of creationism is simply the claim that God created the world in a single event; Gosse

---

[9] Leplin (1982) develops the idea that an ad hoc hypothesis that serves to repair a defect in some theory is one that reconciles the theory to the anomaly but in some sense "does not go far enough" in redressing the theory's problems, an idea that Leplin also discusses in his (1975). Leplin held in these papers that scientists sometimes judge that certain anomalies faced by a theory call for substantive rather than piecemeal reform of the theory – but an ad hoc hypothesis offers only piecemeal reform.

[10] Cf. Worrall (1985), (1989), but especially his (2002) and (2005). See also Scerri and Worrall (2001).

proposed to reconcile the core to the fossil record by adopting the auxiliary assumption about God's motives noted above. The fossil record is thus accommodated by the 'Gosse dodge.' Worrall explains that the support provided by the fossil record to the modified creationist theory is merely conditional: if there is independent support for the core idea of creationism, then the fossil record supports the modified theory. Absent such independent support, the fossil record is powerless to support it. A genuine prediction made by creationist theory would be one that – if verified – genuinely supported the core and would thus supply unconditional support for the modified theory.

Worrall emphasizes that his position accords no actual significance to human motivations – his account is a strictly logical one which ultimately focuses on just the relationship between core, auxiliary claims, and evidence (his account is thus one I would dub a 'weak' account of predictivism, cf. below). Rather than embark on further discussion of Worrall's theory here, however, I refer the reader to Barnes (2005) in which I discuss and criticize Worrall's account in some detail – and see Worrall's (2005) reply. I will offer further discussion of Worrall's account in Chapter 3 in connection with the story of Mendeleev's predictions.

1.4.5 Genetic theories of predictivism

A genetic account of predictivism is one that holds that predictivism is true because successful prediction, unlike accommodation, constitutes evidence that the theory was generated by a dependable source. Genetic theories thus posit a kind of 'confirmational rebound effect': the success of the theory's prediction(s) provides support to the claim that the genesis of the theory was trustworthy, but this support then rebounds off the genetic process and is reflected back to the theory itself. I will discuss two important examples of genetic theories, those of Maher (1988) and Kahn, Landsberg and Stockman (1992). (The primary aim of the next three chapters of this book will be to develop a new genetic theory of predictivism.)

Maher's (1988) presents the *Coin flip* example presented above. To review: in the first scenario, a subject (the accommodator) is presented with E, which is the outcome of a sequence of 99 flips of a single fair coin. E forms an apparently random sequence of heads and tails. The accommodator is then instructed to tell us the outcome of the first 100 flips – he responds by reciting E and then adding the prediction that the 100th toss will be heads – the conjunction of E and this prediction about the last toss is T. In the other scenario, the subject (the predictor) is asked to predict the first 100 flip outcomes without witnessing any outcomes – the predictor endorses theory T. Thereafter the coin is flipped 99 times, E is established, and the predictor's first 99 predictions are confirmed. It seems the only reasonable position is that T is substantially better confirmed in the predictor's scenario than in the accommodator's, despite the fact that it would seem to be just evidence E that is offered in support of T in each case. If we allow 'O' to assert that evidence E was input into the construction of T, predictivism asserts:

$$P(T/E \sim O) > P(T/EO) \qquad (1)$$

Maher argues that the successful prediction of the initial 99 flips constitutes persuasive evidence that the predictor 'has a reliable method' for making predictions of coin flip outcomes. Once this is established, we are prone to believe the predictor's prediction about the one hundredth flip, for there seems to be no reason not to extend our belief in the predictor's method's reliability to the 100th flip. T's consistency with E in the case of the accommodator provides no particular evidence that the accommodator's method of prediction is reliable – thus we have no particular reason to endorse his prediction about the 100th flip. Allowing R to assert that the method in question is reliable, and M_T that method M generated hypothesis T, this amounts to

$$P(R/M_T E \sim O) > P(R/M_T EO) \qquad (2)$$

Maher's (1988) provides a rigorous proof of (2), which in turn is shown to entail (1) on various straightforward assumptions.[11]

Maher is deliberately non-committal about the form the predictor's putatively reliable method might take – it might consist of the derivation of coin flip outcomes from some true mechanical theory applied to the dynamics of a coin flipped from some carefully observed initial position with some particular force, or consist of the predictor's consultation of a crystal ball that never fails to generate true predictions about coin flip outcomes. The predictor may be reading his predictions off a chart that for some reason happens to reliably predict coin flip outcomes, or perhaps the predictor is simply clairvoyant. Maher's basic point would seem to be this: the more likely we regard the possibility of a predictor's finding some

---

[11] For criticisms of Maher's analysis see Howson and Franklin (1991) and Maher's (1993) reply. For further discussion of Maher's program see Barnes (1996a) and (1996b) (the latter presents the arguments developed in Chapter 5 of this book in terms of Maher's theory of predictivism). See also Achinstein (2001: 217–221) and Harker (2006).

reliable method of coin flip prediction, the more willing we are to conclude that the predictor's 99 true predictions are the result of his possessing a reliable predictive method. The point is that the inductive strength of the inference to such reliability is tied to a background assumption regarding the accessibility of reliable methods to the predictor, an assumption which establishes the prior probability that any particular method of prediction selected from those in use in the scientific community is reliable (i.e., P(R) for that method).

Maher begins with the simplifying assumption that any method of prediction used by a predictor is either completely reliable (this is the claim abbreviated as 'R') or is no better than a random method (~R). (Maher (1990) shows that this assumption can be surrendered and the argument that predictivism holds (under certain conditions) remains sound even while allowing that there is a continuum of degrees of reliability of scientific methods.) Taking the claim that the predictor's method led to the prediction of E as background knowledge, we have by Bayes' theorem:

$$P(R/E) = \frac{P(R)P(E/R)}{P(E/R)P(R) + P(E/\sim R)P(\sim R)} \qquad (3)$$

But given the stipulated background knowledge, R entails E, so $P(E/R) = 1$, and $P(\sim R) = 1 - P(R)$, so

$$P(R/E) = \frac{P(R)}{P(R) + P(E/\sim R)(1 - P(R))} \qquad (4)$$

The point of (4) is to reveal what ultimately drives the inference to the reliability of the predictor's method: put in qualitative terms, where M has already predicted E, we are prone to judge that M is reliable on E because we hold that it is substantially more likely that the method that generated E (whatever it is) is reliable (a probability measured by P(R)) than that E just happened to turn out true though R was no better than a random method (a probability measured by $P(E/\sim R)(1 - P(R))$). Thus, when presented with the case of the predictor in Maher's example, where we assume an approximately fair coin, we simply judge that we are more likely to stumble on a subject using a reliable method M of coin flip prediction that we are to stumble on a sequence of 99 true flip predictions that were merely lucky guesses (so $P(R) >> (1/2)^{99}(1 - P(R))$).[12]

---

[12] It should be emphasized that the summary of Maher's reasoning I provide here is a greatly simplified version of the sophisticated and mathematically complex proofs presented in Maher's (1988).

Another genetic theory of predictivism has been propounded by Kahn, Landsberg and Stockman (hereafter, KLS) in their (1992). Where Maher posits predictive methods of unequal reliability, KLS posit scientists of unequal 'talent', where a scientist's talent is a measure of the likelihood that a theory that a scientist proposes is a 'good' theory (in one of several senses, cf. p. 505). Where Maher idealizes that any method used is either wholly reliable or is no better than a random method, KLS idealize that there are just two types of scientists: type *i* scientists are more talented than type *j* scientists. This holds, KLS assume, whether or not they are 'looking first' (viz., accommodating evidence in the construction of a theory) or 'theorizing first' (viz., proposing the theory and then checking its empirical consequences). The basic point is that when a scientist theorizes first and makes a prediction that is subsequently confirmed, this raises the probability that the scientist is an *i*-type rather than a *j*-type scientist, thus raising the probability that the proposed theory is true more than if the scientist had looked first and constructed the theory with the acquired evidence in hand (here again is the rebound effect). The analogy between this analysis and Maher's is clear: successful prediction, for KLS, raises the probability that a scientist is talented, but KLS's very broad sense of talent is similar to Maher's very broad sense of 'using a reliable method': both Maher and KLS argue for a high credibility for theories when they are proposed by scientists who have demonstrated predictive skill. They both embody the genetic approach to predictivism, as they accept that prediction, not accommodation, testifies to the credibility of the theory-producing entity.[13]

White (2003) presents an account of predictivism that falls roughly in this same category. White argues that the fact that a theorist selects a novelly successful theory provides evidence that the theorist "was reliably aiming at the truth" (663). If the theorist selects a theory simply because it fits the given data this provides no particular evidence of such reliable aiming. White's theory, like Maher's, emphasizes that the relevance of prediction in theory assessment consists of the information it provides about the conductivity of the theorist's scientific method to truth (or at least predictive accuracy).

I will consider the merits of Maher's, KLS's, and White's genetic accounts of predictivism in Chapter 3 and compare these accounts to my

---

[13] KLS (1992 Section 3) also sketch a social planning scenario in which a scientist's choice to accommodate or attempt to predict data carries information about the scientist's level of talent – I argue in Chapter 5 that their analysis of this point suffers from its failure to appreciate the relevance of the size of the predicting community.

own genetic theory of predictivism. For the moment, I want to consider the relationship between genetic accounts and the Keynesian dissolution of predictivism discussed in the previous section. The gist of the Keynesian dissolution, once again, is that theories that are proposed without being built to fit the data that supports them are typically proposed because there is some other preliminary evidence for them – when the empirical consequences of such theories are subsequently confirmed, then, such theories are confirmed by two independent sources: the preliminary motivating evidence and the confirmed empirical consequences. But a theory that is merely built to fit the data presumably has no independent motivation (else it was not *merely* built to fit the data), and thus is confirmed by the data alone. There is a certain resemblance between this analysis and the genetic accounts of Maher and KLS. For the fact that a theory is proposed without being built to fit the data testifies on both accounts to the existence of a certain preliminary evidence for the theory – or so one might argue. For Maher, there is some preliminary non-negligible reason to think that the predictions were produced 'by a reliable method.' It is not particularly clear what this amounts to, but perhaps the fact that predictions followed from a theory for which there was some preliminary evidence could instantiate the concept of a 'reliable method' (though clearly, such a 'method' would be absolutely reliable only if the preliminary evidence constituted absolute proof of the theory). Likewise, for KLS, the fact that a theory is proposed by 'theorizing first' contains a form of preliminary evidence that is measured by the prior probability that the scientist is talented – the concept of a 'talented scientist' is likewise far from perspicuous, but perhaps a talented scientist is one who is prone to propose theories for which there is considerable preliminary evidence.

Despite these possible similarities, there are deep differences between genetic accounts and the Keynesian dissolution. For one thing, the Keynesian dissolution entails that the relationship between the demonstrated truth of empirical consequences and the preliminary non-consequential support of the theory is merely one of combination – the two sources of support combine to produce an overall greater support for the theory than in the case of accommodation. By contrast, the genetic accounts presented here argue that the relationship between predictive success and the assessed credibility of the genetic process is revisionary: successful prediction revises our assessment of the credibility of the preliminary theory-generating process. These issues will receive further discussion in Chapter 3.

A still more obvious difference between the Keynesian dissolution and the genetic accounts is that the Keynesian distinction serves to repudiate predictivism, while the genetic accounts of course aim to vindicate it. Some would argue, however, that the relationship between pro-predictivist and anti-predictivist positions is not necessarily as antagonistic as it might seem. We will consider this issue below.

1.5 WEAK AND STRONG PREDICTIVISM

Sober and Hitchcock (2004) posit a helpful distinction between 'strong' and 'weak' predictivism.[14] Strong predictivism holds that prediction is in and of itself superior to accommodation. A theory that predicts some phenomenon is, for this reason and no other, more strongly confirmed than one that accommodates it. Weak predictivism holds that a theory that successfully predicts a phenomenon is typically or at least often possessed of some other feature that confers favorable epistemic status on it, whereas this connection tends not to exist in the case of accommodation. It is the presence or absence of this other feature that is epistemically relevant, while successful prediction and accommodation play merely symptomatic roles.

Strong predictivism, thus defined, holds that the fact that a theory was (or was not) built to fit the data is by itself epistemically relevant. An example of a strong predictivist position is the position that holds that the empirical facts that a theory was built to fit cannot confirm the theory, because the theory construction process guarantees that the theory stood no chance of being refuted by the facts (e.g. Giere 1983: 161). Strong predictivists hold that the knowledge of certain facts about how a theory was built is essential to a complete understanding of the theory's epistemic standing. Thus strong predictivism fully embraces biographicalism. Weak predictivism allows that a full understanding of such epistemic standing need not include such knowledge. Presumably a proponent of the Keynesian dissolution could re-cast her position as a defense of weak predictivism (rather than as an attack on predictivism) by pointing out that successful prediction, but not accommodation, tends to be correlated with the existence of independent non-consequential evidence for the relevant theory. The distinction between a weak predictivist position and an anti-predictivist position seems, at first glance, primarily rhetorical: both reject biographicalism.

Sober and Hitchcock bill their own (2004) account as a weak account – they claim that under certain conditions theories that are built by the process of data accommodation are prone to be the product of a fallacious mode of theory construction – one known as 'overfitting the data.' When

[14] The distinction is also drawn in Lange (2001) and White (2003) in a roughly similar way.

the data scientists seek to explain is 'noisy' (i.e. when such data is the product of an imperfect observational process) the best theory will not be one that perfectly fits the data. Theories that successfully predict data are systematically less likely to be the product of overfitting (for one cannot adjust a theory to perfectly fit data that is not yet possessed). But ultimately it is the presence or absence of overfitting (i.e. the 'too close' relationship between theory and data) that is epistemically relevant to theory assessment, not the issue of whether the theory was built to fit the data. Lange (2001) could likewise be classed as a proponent of weak predictivism, as he claims that theories that are the product of many accommodations are statistically more likely to be 'arbitrary conjunctions' which are less susceptible to confirmation than theories which enjoy predictive success. Such weak predictivisms seem to offer a compromising solution to the paradox of predictivism: they retain a watered down commitment to the relevance of prediction, but dispense with the apparent absurdity of biographicalism.

However, weak predictivism is an ambiguous position. A weak predictivist could maintain that, insofar as prediction and accommodation are merely correlated with the presence or absence of some underlying feature F that carries actual epistemic significance, actual scientists who assess the credibility of theories pay no attention to facts about prediction or accommodation and focus entirely on the presence or absence of F. But a weak predictivism could also maintain that while it is only the presence or absence of F that ultimately matters, it is nonetheless the case that facts about prediction and accommodation can be epistemically significant precisely in virtue of their symptomatic roles. This could be the case if theory evaluation were carried out from a standpoint in which it was difficult or even *de facto* impossible for the scientist who evaluates a theory's probability to determine whether F is present, but in which it was possible to determine whether data was predicted or accommodated (see Lipton 2004: 177–183).

I propose to label as a 'tempered predictivism' any weak predictivism that emphasizes that facts about prediction and accommodation can carry actual epistemic relevance for evaluators in virtue of their symptomatic roles. Any theory of weak predictivism which links prediction to some feature F can in principle be reconstrued as a theory of tempered predictivism insofar as actual theory evaluators rely on prediction and accommodation facts in their attempt to determine whether F is present. A theory of tempered predictivism is thus essentially a genetic theory of predictivism. Sober and Hitchcock, e.g., make something like this point about their theory of predictivism, pointing out that in cases in which it is

difficult to determine whether theories were constructed by overfitting the
data, the fact that data was predicted is evidence of the absence of overfitting, and successful prediction is evidence that overfitting did not occur.
Thus successful prediction serves as an indication – a symptom – of
dependable theory generation (2004: 4). Likewise Lange suggests that the
fact that a sequence of phenomena were predicted (rather than accommodated) could constitute evidence that the predictor was led to posit this
sequence by way of something we have not noticed that ties the sequence
together and renders them a 'non-arbitrary conjunction' (2001: 581). Weak
theories of predictivism, in general, become genetic theories when cast in
tempered form. A weak predictivism which denies even any symptomatic
(viz., epistemic) role to prediction and accommodation in theory evaluation and merely notes the correlation between prediction and some feature
F I will call 'thin predictivism.' Thin predictivism thus holds that prediction is correlated with F but denies that actual scientists accord any
epistemic significance to prediction itself. For thin predictivism, prediction
and accommodation have no genuinely epistemic or symptomatic role to
play in actual theory evaluation.

At first glance, Maher's method-based account of predictivism seems to
be a strong predictivism – for it clearly endorses the epistemic significance
of facts about theory construction. But closer inspection reveals another
story: Maher concedes that his proposed species of predictivism will fail in
cases in which the theory evaluator knows how reliable the operative
method of prediction is – for of course his basic claim is that predictivism
holds true insofar as predictive success functions as evidence for method
reliability (1988: 277). Thus Maher concedes that biographical information
is eliminable, but nonetheless often epistemically important given its
symptomatic role. Maher's predictivism is not strong but tempered. The
same holds of KLS's account, as they emphasize that predictivism will fail
in cases in which information about the degree to which individual
scientists are 'talented' is public information – for predictivism holds for
KLS insofar as predictive success functions as evidence for talent.[15]

To take stock: strong predictivism is endorsed by philosophers who take
facts about the theory construction process to be essential to a full assessment of a theory's credibility. I will argue in Chapter 4 that primary

---

[15] Interestingly, while KLS endorse a tempered predictivism, they nonetheless offer a picture of theory
evaluation which is strongly biographicalist, insofar as facts about the mental lives of scientists
(specifically their degree of 'talent') is apparently essential for a full assessment of theory probability.
Thus their position reveals that biographicalism is a necessary but not sufficient condition for strong
predictivism.

argument for strong predictivism (instanced, for example, by realist analysis like Giere's noted above) is fallacious; I will conclude that strong
predictivism is a defunct position. Weak predictivism, as it has turned out,
comes in both tempered and thin forms. It is tempting to equate thin
predictivism with anti-predictivism, for it denies that prediction and
accommodation facts play any role in actual theory assessment. But of
course, a theory of thin predictivism could nonetheless be illuminating, if
the underlying feature F it identifies is both surprising and plausible. Hope
for a robust predictivist thesis, in my view, lies with tempered predictivism,
which avoids a thoroughgoing biographicalism while conceding the epistemic significance of prediction and accommodation in their symptomatic roles. Tempered predictivism, as we have seen, applies to theory
evaluation that takes place in a context in which theory evaluators are less
than perfectly informed of the total body of evidential information that
bears on theories. It is an interesting and by no means simple question to
what extent actual theory evaluation takes place by imperfectly informed
evaluators. We turn to this question in the next chapter.

1.6 OVERVIEW OF THE ARGUMENT OF THIS BOOK

I hope the reader is impressed at this point with the variety of positions that
contemporary philosophers have adopted on the subject of predictivism.
As of the early twenty-first century, the questions whether predictivism is
true and how it should be understood remain mired in controversy. My
sense is that the paradox of predictivism is still with us today.

Why is the paradox so hard to resolve? One explanation, I shall argue, is
that many philosophers of science today possess a particular deeply held
intuition about the methodology of science – this is the assumption of
epistemic individualism. Epistemic individualism holds that a rational
agent, when determining how credible a particular theory is, relies entirely
on her own judgment in assessing the evidence that bears on the theory.
Epistemic individualists, that is, do not defer to the judgments of other
agents about some theory T's probability when assessing T themselves.
Epistemic individualism seems to many as though it must characterize
the deliberations of expert scientists, the sort of agent whose deliberations
the philosophy of science presumably purports to understand. I will argue
in Chapter 2, however, that the widespread belief in epistemic individualism amounts to a pervasive myth about theory evaluation. Theory
evaluation – even at the expert level – is characterized to a significant
degree by epistemic pluralism. An epistemic pluralist is a theory evaluator

who counts the judgments of other agents about the probability of T as epistemically relevant to her own assessment of T. Showing this is the primary task of Chapter 2. Another purpose of Chapter 2 is to sketch, in informal terms, the account of predictivism (including a new account of 'novelty') that will be developed in subsequent chapters, an account which depends in large part on the recognition of the reality of epistemic pluralism. One consequence of this account is that biographicalism is transformed from an absurdity into a common sense truth about theory evaluation.

In Chapter 3 I propose a rigorous theory of predictivism which fully works out the proposal sketched in Chapter 2. Predictivism comes in two forms, which correspond to two distinct attitudes toward the theorist (whether predictor or accommodator). One of these, unvirtuous predictivism, is based on a certain skepticism about the logical trustworthiness of agents who endorse theories. Prediction, unlike accommodation, can alleviate such skepticism. The other form, virtuous predictivism, takes theory endorsing agents to be logically impeccable but locates the truth of predictivism in the greater propensity of predicted evidence to confirm the background beliefs of the predictor. Both virtuous and unvirtuous predictivism come in thin and tempered forms. A detailed application of this account of predictivism is provided based on Mendeleev's proposed periodic law of the elements. I compare my account of this episode with competing accounts and argue that mine is superior.

Chapter 4 turns to a topic that is absolutely fundamental for present purposes: the realist/anti-realist debate. Realism – a position which regards the theories of the 'mature sciences' as at least approximately true – has been defended by the claim that realism is the only position which does not make the vast empirical successes of the mature sciences a miracle. But there is a long-standing tradition among realists that the sorts of true empirical consequences that count as serious evidence for the truth of theories are not all their true consequences but the true consequences such theories were not built to fit. Many realists in this tradition have endorsed a particular argument for predictivism which runs roughly as follows: where T entails true empirical consequences that it was not built to fit, we ought to infer that T is true because the truth of T is the only viable explanation of this kind of empirical success. To avoid postulating the truth of T is to render such success a mere coincidence, and this is an intolerable conclusion. This argument may be implicit in Whewell (cf. ft. 4) but is explicitly endorsed by many others. I argue there that this argument is straightforwardly fallacious both as an argument for realism and as an explication of predictivism. The project of providing a non-fallacious argument for realism in which novelty plays a prominent role will lead us back to the theory of predictivism I developed in Chapters 2 and 3. I go on to show how the non-fallacious argument for realism can be defended against certain objections. The import of these defenses is that predictivism turns out to be a doctrine that is available only to the realist. I consider also a seminal paper by Magnus and Callendar (2004) and show how their critique of the realist/anti-realist debate can be answered by the various points developed in this chapter.

In the course of the discussion of the realist/anti-realist debate a particular claim emerges as particularly fundamental, and this is the claim that novelly successful theories emerge at a rate that is too high to be attributed entirely to chance. Whether this claim (which I dub the 'high frequency assumption') is justified or not has much to do with the outcome of the realist/anti-realist debate. But the high frequency assumption brings to light a fundamental aspect of novel confirmation, and that is the relationship between a particular novel success concerning a particular empirical claim and the number of agents making predictions about related empirical claims. Intuitively, the more predictors there are, the greater the probability that some agent will make successful predictions by chance. The epistemic significance of a novel confirmation thus cannot be assessed by considering the various facts about the predictor herself, but require assessing various facts about the 'predicting community' in which she is located – a phenomenon I refer to as 'social predictivism.' These matters are explored in detail in Chapter 5.

As noted, I argue in Chapter 2 that epistemic individualism is a myth about theory evaluation in science that it is high time to renounce. One of the reasons for the popularity of this myth is that there is a long standing tendency to imagine that the genuinely rational individual is a suitably skeptical person who does not rely on sources of information other than her own ability to assess the relevant evidence. On such a view, the epistemic pluralist seems guilty of being insufficiently skeptical to qualify as rational. But this view can be refuted by showing how epistemic pluralism can be formulated in such a way as to meet stringent criteria of rationality. In Chapter 6 I provide a rigorous conception of how the epistemic pluralist incorporates information about the trustworthiness of other agents in a way that is straightforwardly rational. Thus epistemic pluralism emerges as a rational belief forming process, a conclusion importantly related to my claim that predictivism can be itself a rational method for theory evaluation.

In Chapter 7 I turn to an issue which has long been entangled with the issue of predictivism, and this is the problem of old evidence (originally noted in Glymour (1980)). Suppose E is intuitively evidence for H and H entails E. If evidence E is already known (and thus 'old evidence'), so that $p(E) = 1$, it follows immediately from Bayes' theorem that $p(H/E) = p(H)$. But this equality suggests that E does not confirm H at all if we accept that $p(H/E) > p(H)$ iff E is evidence for H. Prima facie, the problem of old evidence provides a simple justification for the null support hypothesis (the Popperian claim that old evidence cannot support a hypothesis at all). But the null support hypothesis is intuitively absurd (as Mayo's example of the SAT scores indicates), suggesting that there is something amiss with the Bayesian reasoning sketched above. A detailed analysis of the old evidence problem and its multiple connections to predictivism will be provided in Chapter 7. The old evidence problem, I will conclude, neither supports nor undermines the reality of predictivism in theory evaluation – despite initially plausible reasons to the contrary.

Chapter 8 offers a summary of the various claims of this book.

CHAPTER 2

Epistemic pluralism

2.1 INTRODUCTION

John is a bookie who posts odds on sporting events. In determining the probability that the Dallas Cowboys will defeat their next opponents, John takes into account the various strengths and weaknesses of both teams. John then declares that this probability is 0.5. But at this point John learns that another bookie, Trish, has declared that the probability of a Cowboy victory is 0.01 – John is surprised, for he holds Trish's opinions in such matters in high regard. John may feel that he should modify his own posted probability on the basis of knowing the probability that Trish has posted. But how, exactly, should he go about assimilating this information? John's predicament raises the general problem of how rational agents do or should make use of other persons' posted probabilities as evidence. How, if at all, is John to combine the content of his own deliberations about the result of the Cowboys' game with the information about Trish's probability to compute an updated probability?

Philosophers have differed over whether scientists (and rational agents generally) do or ought to consider other persons' probabilities as epistemically significant. Earman argues (1993: 30) that "It is fundamental to science that opinions be evidence-driven" rather than driven by reference to other persons' opinions. But Foley (2001: 97–107) makes the point that, for any rational agent, refusing epistemic significance to the judgments of others is incoherent when one holds one's own judgments to possess such significance – if I can have reason to believe that my own judgment is reliable, I can have reason as well that my peer's judgment is reliable – or more reliable – than my own, and my own assessment of the world must respect this fact.

It is easy to understand why such controversy exists. In his essay "Epistemic Dependence," John Hardwig argues that there is a long standing tradition in the Western world of 'epistemic individualism,' which maintains that one may not rationally defer to the opinions of others in forming beliefs. Hardwig diagnoses the causes of this tradition as follows:

The idea behind [epistemic individualism] lies at the heart of one model of what it means to be an intellectually responsible and rational person, a model which is nicely captured by Kant's statement that one of the three basic rules or maxims for avoiding error is to 'think for oneself'. This is, I think, an extremely pervasive model of rationality – it underlines Descartes' methodological doubt; it is implicit in most epistemologies; it colours the way we have thought about knowledge. On this view, the very core of rationality consists in preserving and adhering to one's own independent judgment. (1985: 340)

Hardwig ultimately concedes, however, that such a model provides a 'romantic ideal' which ultimately results in less rational belief. Once we realize that the reliability of other persons as cognitive agents is something that can be scrutinized and evaluated critically, there is no reason not to avail ourselves of the evidential significance of other persons' beliefs – including their posted probabilities. But we should be aware that this proposal strikes at the heart of much of what has long been taken for granted about reasonable belief.

   An 'epistemic pluralist' is an agent who counts the judgments of some other agents as a form of evidence about the world. In this chapter I will argue that pluralist theory evaluation pervades the practice of science in multiple ways. Before proceeding it is probably worth clarifying what this argument does and does not involve. For example, in defending what I call epistemic pluralism I am not entering the domain of the now large epistemological literature on testimony. Beginning with Coady (1992), this literature treats testimony as a fundamental source of information about the world, on a par with sensation, reason and memory, and attempts to determine the role of testimony in a comprehensive epistemological theory of how human beings acquire knowledge (cf., e.g., Lackey and Sosa 2006). My project is more mundane. I take for granted that human beings are capable of acquiring many kinds of knowledge, or at least empirically adequate beliefs. My thesis is that working scientists make important use of the judgments of other scientists in their evaluations of scientific theories. Epistemic pluralism should not, furthermore, be confused with 'communitarian epistemology' which claims that the knowing subject is not primarily the individual but the community (cf. Kusch 2002). Epistemic pluralism retains individuals in the role of knowers but emphasizes their epistemic dependence on other agents.

   Epistemic pluralism is a result of the fact that scientists are widely compelled to evaluate theories while not themselves in possession of all evidentially relevant information. Thus, I will argue, pluralist evaluators are prone to adopt a tempered predictivism in their acts of theory assessment. This is because pluralist evaluators are in need of information about the credibility of the various agents they encounter, and because an agent's predictive successes (or lack of them) can provide such information. This is perhaps the most straightforward account of predictivism it is possible to formulate: agents who endorse theories that go on to be novelly successful tend to be agents who know what they are talking about. Thus novel success testifies to the general credibility of the agent, and such credibility rebounds off the agent and back to the theory the agent endorsed. An accurate understanding of predictivism in theory evaluation can occur only in the context of a broader understanding of the role of epistemic pluralism in the evaluation of scientific theory. To establish that pluralism pervades theory evaluation in science is by itself sufficient to suggest, at least, the pervasiveness of tempered predictivism.

Before proceeding it bears noting that the move to pluralism clearly entails that biographicalism will be viewed in a more positive light. Pluralist evaluators will not find biographicalism counterintuitive at all, for information about the 'life stories' of agents will possess obvious epistemic relevance from their perspective. Agents who are known to possess extensive training in their fields will tend to be granted more authority than others, for example. Moreover, as I will argue below, pluralists will not find information about which facts were used by theorists irrelevant either (in a way that will be explained in the next section). Pluralism promises to make both predictivism (in its tempered form) and biographicalism into common sense truths – and thus promises a welcome solution to the paradox of predictivism.

In section 2.2 I argue that the notion of novelty which we have applied thus far – E is a novel confirmation for T only if T is not built to fit E – is inadequate for the purposes of a pluralism-based predictivism. In section 2.3 I develop a model of pluralist theory evaluation that is clearly usable by non-experts, and in section 2.4 I argue for the prevalence of expert pluralism. This will pave the way for a new theory of tempered predictivism, sketched in this chapter but developed and illustrated in much more detail in the next chapter. In Chapter 6 I will return to the topic of epistemic pluralism and develop a rigorous account of pluralist theory evaluation.

2.2 NOVELTY FOR PLURALIST EVALUATORS

In the previous chapter we considered briefly the various accounts of novelty that have been proposed, but the nature of novelty has been the

subject of a vast philosophical literature which we have by no means fully
reviewed. Interested readers are directed to Leplin (1997 Ch. 2) for excellent
discussion. Thus far we have focused on a single conception according to
which E is a novel confirmation for T if and only if T was not built to fit E.
This is, these days, the most widely used conception of novelty and it is
standardly deemed 'use-novelty' – because it makes E novel for T when E
was not 'used' in the construction of T.[1] Let us call any agent who assigns a
probability to some theory T a 'T-evaluator.' A T-evaluator is a pluralist
insofar as she regards one form of evidence pertinent to T to be the
probabilities assigned by other agents to T. A T-evaluator is an individualist if she does not do this but considers only the relevant data and her
own scientific knowledge as relevant to her judgment about T's probability. How intuitively plausible predictivism appears depends, as suggested
above, on whether one thinks of theory evaluators as pluralists or as
individualists. For consider a theorist who constructs a theory T which
entails a true piece of evidence E, and another agent who plays the role of
evaluator. Suppose the evaluator is a pluralist. If the theorist simply built T
to fit this true consequence, this fact (that the theorist constructed a truth
entailing theory) can be explained adequately by noting how the theory was
built. But suppose that T was not built to fit the truth – how is this fact to
be explained? For our pluralist evaluator the best explanation may be – as
Maher would argue – that the theorist had some reliable method for
constructing T – for if this were not the case, the construction of a truth
entailing theory was just a lucky accident.[2] So the evaluator is driven to
the hypothesis that the constructed theory is probably true (or at least

---

[1] The premium on use-novelty was anticipated by Popper's prohibition on ad hoc hypotheses – such
hypotheses were disreputable because they were built to fit some known data. This premium has been
endorsed by Giere (1983, 1984: 159–161), Redhead (1986: 117), Maher (1988, 1990, 1993), Zahar (1973,
1989: 14–15), Kahn, Landsberg and Stockman (1992), Barnes (1996a, 1996b), and Psillos (1999).
Musgrave (1988) cites many proponents of the miracle argument for realism that essentially endorse
the premium on use-novelty, such as Whewell, Duhem (despite his ultimate anti-realist stance), and
Clavius. Many philosophers (such as those discussed in Chapter 1) who either deny predictivism or
defend it only in its weak form (like Sober and Hitchcock, and Lange) nonetheless make use of the
'use-novelty' conception. Worrall appears to endorse the use-novelty conception (e.g., his (2002)) but
explains in other work that the facts about how a theory was constructed can be given a logical rather
than psychological explication (1985, 1989, 2005). Thus it seems to matter, for Worrall, not whether a
theorist did in fact use some piece of evidence in constructing a theory, but whether such evidence was
needed to construct the theory, a question he thinks can be settled by inspecting scientific literature in
the public domain.

[2] Recall that Maher (1988, 1990) claims that predictive success is evidence that the predictor has a
reliable method of making predictions – in his (1993) he notes that one form of a reliable predictive
method is a reliable method of constructing theories that entail true predictions.

empirically adequate). The evaluator's reasoning is pluralistic insofar as it critically involves a certain element of trust in the theorist. She judges that it is more probable that his basis for constructing T is reliable than that the construction of a truth-entailing theory would occur by chance – thus the epistemic significance of use-novelty for pluralist evaluators as Maher explains it. But apart from other problems with Maher's analysis that will be discussed in Chapter 3, there is something misguided in spelling out the relevant epistemics in terms of the act of theory construction. For to construct a theory, presumably, is simply to formulate the theory by a cognitive act of the imagination. And it is a mistake, I believe, to claim that such an act is what actually carries epistemic import. Consider, for example, that some competent scientist Connie constructs a new theory T that entails some (otherwise unlikely) empirical consequence N_1 that is subsequently shown to be true. A pluralist evaluator is impressed and is prepared to give high marks to T on the basis of the reasoning given above. But suppose that at this point Connie insists that while she did construct T, she by no means endorses T – in fact, she is certain that T is false. She claims that if T is tested again, now by testing for another novel (also very unlikely) consequence N_2, T will be shown false. There is no question about her sincerity. Now at this point I suspect that the success of N_1 would testify to Connie's credibility as a predictor for the evaluator – but for this very reason, it would no longer constitute strong evidence for T despite the fact that N_1 was a use-novel confirmation of T. This is because Connie did not endorse T upon constructing it. Now let us suppose that another competent scientist, Endora, appears and claims that Connie is wrong. Endora endorses T and claims that it is probably true – and that N_2 will be shown to be true when tested. Now suppose that N_2 is shown to be true. The truth of N_2 could testify, for the evaluator, to the credibility of Endora rather than Connie – specifically, to Endora's possession of a truth-conducive basis for endorsing T – despite the fact that she did not construct T. This suggests that the epistemically relevant act is not construction but endorsement. The long tradition of spelling out novelty in terms of the notion of construction reflects, perhaps, an uncritical (if understandable) tendency to assume that any constructor of a theory also counts as an endorser of that theory.

What does it mean to endorse a theory? To endorse a theory, I claim, is in some way to affirm the theory in the presence of some evaluator(s) (other than the endorser himself, who is obviously also an evaluator). This would consist in posting a probability for the theory that is either high or at least not so low as to constitute a primarily skeptical or noncommittal attitude toward the theory. Obviously, there are many degrees of endorsement. In most cases of public endorsement scientists provide their evaluators with some sense of how probable they take the theory to be. For example, one important location in which theories are endorsed is in journal articles – such articles typically have a 'Conclusion' section toward the end in which the implications for the data are drawn for various relevant theories – such theories as the author may want to endorse are given there along with the author's sense of how strongly the data supports them (Suppe 1998: 403). My proposal is that the act of endorsement is to be thought of as an act by which a scientist displays for some evaluator(s) a probability for the endorsed theory that does not fall below some contextually fixed limit L. I propose two conditions on L: (1) L must be no lower than the evaluator's own probability (if she has one). If L is lower than the evaluator's probability, then the evaluator is likely to view the other agent as having a more skeptical view of T than the evaluator – and thus does not qualify as an endorser vis à vis that evaluator. (2) L must be sufficiently high so that, should some new piece of evidence for T be obtained, the pluralist's evaluation of the theorist's credibility is at least somewhat increased because of the new evidence's confirmation of T.[3] Clearly, if the theorist posts a sufficiently low probability for T, a subsequent novel success for T would fail to increase (and could actually decrease) the credibility of the theorist from the evaluator's standpoint.[4] In sum, the endorsement of a theory is a special case of posting a probability, one in which the posted probability is no lower than L.

I claim that N (a known consequence of T) counts as a novel confirmation of T relative to agent X insofar as X posts an endorsement-level probability for T (i.e. a probability greater than or equal to L) that is based on a body of evidence that does not include observation-based evidence for N. That is, X does not base her endorsement of T on any

---

[3] It is important that the increased credibility of the T-endorser be attributable to N's confirmation of T because it is possible that an agent could post a low probability for T but endorse some other theory T* that entails N – thus the confirmation of N could increase the credibility of the agent, but not because N confirmed T.

[4] A special case could occur when the evaluator is antecedently certain of the theorist's absolute credibility (viz., infallibility) – thus the confirmation of N cannot increase the probability of credibility. My position is that rational evaluators will never be absolutely certain of a theorist's infallibility – thus there should always be a little room for credibility enhancement. This squares with the general policy of not assigning probabilities 0 or 1 to contingent statements (which I follow throughout this book).

observations of N's truth. Confirmations that are novel in this sense I deem 'endorsement-novel.'[5] Predictivism now amounts to the claim that when true evidence N confirms T, endorsed by X, T is more strongly confirmed (for some evaluator) when N is endorsement-novel relative to X than when it is not.[6] One advantage of endorsement-novelty over use-novelty, as we will see below, is that the notion of endorsement novelty clearly entails that the status of an agent as a predictor is a matter of degree – the more strongly one endorses a theory T, the more strongly one predicts that theory's empirical consequences. On the notion of use-novelty, by contrast, one is either a predictor of a theory's consequences or one isn't – for one either used the relevant piece of evidence in constructing the theory or one didn't. Another advantage is that the traditional distinction between the context of discovery and the context of justification is preserved, a distinction that proponents of use-novelty have been compelled to renounce (such as

---

[5] The conception of novelty elaborated by Leplin (1997 Ch. 3) bears some resemblance to the notion of endorsement novelty defended here. Leplin claims that N is a novel confirmation of T if N is not cited in a rational reconstruction of the reasoning which generated T. It is a condition of adequacy on any such reconstruction that the reconstruction serve to make T 'significant' or 'worthy of consideration.' Thus the reasoning that produced T should not include N but should render T 'endorsable.' However, Leplin's conception of novelty is nonetheless clearly construction based – this is because he needs a conception of novelty that will support his version of the miracle argument for realism, which argues that truth is needed to explain why a theory entails some true consequence if this cannot be explained in terms of how the theory was built. Thus the significance of novelty for Leplin ultimately consists of its role in how the theory was constructed, not on what basis it was endorsed. (We will return to this topic in Chapter 4). As he says in explaining his theory of novelty, "To some extent, the adequacy of a reconstruction is evaluative; the notions of 'worth' and 'significance' to which I appeal carry epistemic value. But at issue here is the provenance of T, not its evaluation once proposed." (1997: 68) Some Lakatosians propose to count N as a novel confirmation for T (and its research program) if N is not used to support another theory, either in the same research program (Frankel 1979: 25)) or in a rival research program (Nunan 1984: 279) (these are variants of the 'theoretical' conception of novelty noted at the beginning of Chapter 1). While there is an emphasis on support rather than construction here, the differences between such proposals and endorsement-novelty are obvious: evidence can be novel for Frankel or Nunan even when it is appealed to in the theorist's preliminary endorsement, and even when the theory is built to fit the evidence. Finally, White, while he explicitly defines prediction and accommodation in terms of whether the theory was designed to fit the evidence (2003: 655), later in the same paper claims that what it means to design a theory to fit some evidence is to select that theory because it fits the evidence. I explicitly proposed the concept of endorsement novelty in Barnes (2005) though the arguments that motivate this concept were articulated in Barnes (2002a).

[6] CUP Referee B queries whether it is a consequence of this proposal that there is no advantage to prediction for a theory that is not already believed (endorsed). It is a mistake, first of all, to equate endorsement with belief – for the endorser may post an endorsement-level probability which falls short of whatever probability is required for belief. My position is that there is no 'advantage' to prediction for a theory that has not been endorsed for the rather trivial reason that there is no prediction apart from an act of endorsement. A prediction of N, on my view, is not a logical entailment but an endorsement – a human act of posting a sufficiently high probability for a theory that entails N (without appeal to observation-based knowledge of N).

Worrall 1985 and Leplin 1987[7]). For on this proposal it is indeed epistemically irrelevant how a theory was constructed – as well it should be. What
matters is on what basis a theory was endorsed.

It is perhaps worth pausing over a possible cause of confusion. Throughout
the discussion above I have assumed a distinction between the endorser – one
who posts an endorsement level probability for T – and an evaluator – one
who assesses the epistemic significance (if any) of this posting. But of course
the endorser is also an evaluator given that he posts a probability for T. I will
typically use the term 'evaluator' to refer to a person other than the endorser
despite this fact. However, as we will see, there are special cases when the
endorser and the evaluator are the same person.

With this new conception of novelty in mind, I will now propose a
general model of pluralist theory evaluation. In the subsequent section, we
consider which scientists can and do make use of this model.

2.3 A PLURALIST MODEL OF THEORY EVALUATION

2.3.1 *Preliminaries*

A scientific community (SC) is a group of individuals whose task it is to
determine the truth about some particular matter having to do with the
natural or social world, and who are in some kind of communication with
one another throughout the deliberative process of searching for this truth.
Although in most of the discussion to follow I will consider SCs of the
usual sort (constituted by individuals with advanced training in the natural
or social sciences) it is important to emphasize that the notion of SC
I adopt is broader than this. For example, a community fitting my
definition of an SC would be a jury whose job it is to determine whether
a particular defendant should be convicted. The jury is presented with a
body of evidence, withdraws so as to consider the evidence, and collectively
determines whether the defendant should be convicted. The SC must, on
the basis of reasoning from other known facts or background belief, argue
for any claim about such guilt or innocence. Other groups which fit my
conception of an SC include scientific communities defined at various
levels of specificity – such as the community of chemists trying to establish
the molecular structure of a particular organic compound, the community
of organic chemists, the community of all chemists, and the entire scientific
community.

---

[7] Leplin (1987) retains a commitment to use-novelty that Leplin (1997) renounces to some degree (cf. ft. 2).

An individual is a member of a particular SC insofar as that individual (1) possesses a body of background belief K which is regarded as a necessary condition for competence in the relevant field, and (2) is involved in some kind of communication with the other members of SC that bears on the project of determining the truth about the germane matter of fact. I refer below to individuals who qualify as a member of a particular SC as 'experts' relative to that SC. This is not to say, of course, that all members of a particular SC have identical background beliefs – as we will see, there are sometimes differences in training or experience that cause certain members to have background beliefs not possessed by others. But in most of these cases there is nonetheless a common core of background belief that is shared by all certified members of the SC – and it is this common core I deem 'K.' These beliefs are called 'background' beliefs in that they are the beliefs that are assumed to be true (at least provisionally) by members of the SC and which serve to guide, often implicitly, the deliberative process by which experts pursue the truth about their domain of expertise.

One of the objectives of the present work is to assess the ways in which non-experts make sense of the judgments of SCs to the effect that particular theories are true, false, or have some intermediate probability. The non-expert does not possess the background beliefs possessed by the experts. This means that the non-expert typically cannot assess the various analyses and arguments that are proffered by the SC or its various members, even if the non-expert were to be exposed to them. The non-expert is presented with only certain types of information. These may include the ultimate judgment of the SC about the theory in question, or indeed whether the SC reached such a judgment. The claim that the SC made a judgment about T suggests that the SC reached either a consensus about T or a near consensus – a failure to make a judgment indicates non-negligible dissensus, as in the case of a hung jury.

2.3.2 *The pluralist model of evaluation applied to SCs by non-experts*

If a non-expert evaluator accords epistemic significance to the judgment of some SC about the probability of some proposition, the non-expert shows herself to be a pluralist. In this case the pluralist regards as epistemically significant not the judgment of some particular individual, but the judgment of an SC. In this section I will develop a model of belief evaluation that may be used (and, I will argue, is often used) by non-experts as they evaluate various claims held by various SCs – I will refer to this model as the pluralist model of theory evaluation. It is important to emphasize from the outset, however, that this pluralist model is relevant not only to the assessment of beliefs held by SCs, but also to the assessment of beliefs by pluralists held by particular members of SCs. The latter point will be taken up below – for the moment we continue to focus on SCs as the relevant entity in the non-expert's evaluation of theory.

When an agent appeals to the opinion of an SC to argue that some proposition p is true, she has applied an argument form that is sometimes called 'appeal to authority' (or 'argumentum ad verecundiam'). There is a considerable body of philosophical literature that discusses this argument form, and the reader is referred to Walton (1997) for an excellent discussion of its history and many logical nuances. As Walton notes, some textbooks of informal logic (in the grip of the appeal of epistemic individualism) classify argumentum ad verecundium as a fallacy. But many philosophers have followed Aristotle in holding that while such arguments are not conclusive or demonstrative, they are not inherently fallacious either. Many authors of informal logic texts have stipulated that appeals to authority are legitimate under certain conditions, such as the legitimate claim of the authority to relevant expertise, the status of the authority as unbiased, whether there is a consensus within the relevant SC, etc.

Not all non-expert trust in so-called scientific opinion should be admired. We live in an age in which it is taken for granted, by many, that science is a trustworthy source of knowledge – and some non-experts defer to science in an uncritical way for this reason alone. But a rational appeal to authority cannot be grounded merely on a passive tendency to accept whatever one's society claims. Furthermore, a rational non-expert will not defer to just any claim that is grounded on appeals to scientific judgment. For example, we are all bombarded by advertisements for products whose merits are 'certified' by appeals to authority that should not be regarded as genuinely persuasive. MacDonald (1985) argues that this is particularly true of food products, which are advertised by endorsements from celebrities, physicians with no training in nutrition, and laboratories funded by the food industry itself (cf. Walton 1997: 6–9). Other appeals to authority are too vague to carry much weight – such as the often heard claim that "many experts agree" that such and such is true. In the absence of specific information about the identity of the experts and the proportion of the relevant SC that agrees with them, such appeals to authority are relatively meaningless. A rational non-expert is someone who is sensitive to these and other conditions on appeal to authority – there is no reason to think that becoming a rational non-expert is easy.

Obviously we cannot do justice here to the myriad of issues that bear on rational appeals to authority. In what follows I will focus on three criteria of pluralistic theory evaluation, which involve (1) reports of particular research projects, (2) consensus and (3) novel confirmation. These three modes of theory evaluation, I will argue, are importantly similar insofar as they are all, under certain conditions and to some degree, epistemically accessible to the non-expert. The non-expert need not grasp the relevant background beliefs to understand that a particular research project has reached some particular judgment – as in the case of a non-expert who, listening to the evening news, hears that an article published in the *New England Journal of Medicine* has suggested that moderate alcohol consumption may help prevent heart disease. Likewise a non-expert may come to understand that there is a consensus of opinion among medical experts that cigarette smoking causes various diseases by learning this from a trustworthy source, such as a well-trained physician. A non-expert may also come to understand that a particular theory endorsed by some SC has undergone a novel confirmation – say by learning that the theory was, after some preliminary endorsement, severely tested and survived the tests.

Clearly, when the non-expert assesses the epistemic significance of some SC's endorsement of some T (either in the form of a report of a particular project or the consensus of the community), it is of the greatest importance to the non-expert how intellectually credible the SC is. Such credibility will be a function of various factors. It will be a function, first of all, of the intellectual credentials of the members of SC. If the non-expert believes them to possess the requisite logical skills and educational background pertinent to the subject matter (such as a group of jurors she knows to be reasonable people who have been presented with all relevant evidence), she will invest SC's endorsement of T with much more significance than if she does not believe this. Such credibility depends as well on her belief that SC is an objective body with no particular reason to make judgments other than those dictated by the evidence alone (as would not be the case if, e.g., SC were constituted by scientists funded by the tobacco industry and T were the claim that smoking causes cancer).

But there are substantive questions to be addressed about how the non-expert may come to know these things. How does a non-member of some SC come to know (or at least reasonably believe) that the members of the SC have the requisite cognitive skills and objectivity to be trusted? My position is that the non-expert must possess a moderately sophisticated body of understanding in order to reach such conclusions. The non-expert must, first, understand on some first-hand basis that there is a distinction between principles of reasoning that are, and are not, truth-conducive. The non-expert may come to appreciate this in a variety of ways – perhaps in the course of acquiring a formal education which nonetheless did not establish his membership in the SC. But he might acquire it in other ways – say by way of a rigorous exposure to the business world, in which he learned from experienced and intelligent business people – or first hand experience – that there are both trustworthy and fallacious modes of reasoning about which business strategies will be successful, etc. Also, the non-expert must appreciate the concept of a community's reward structure – which establishes that community members are rewarded for particular behaviors and not for others. The non-expert in the business world, e.g., will learn that in certain companies employees are rewarded for offering opinions and arguments that can withstand severe scrutiny from the standpoint of the principles of truth-conducive reasoning, while in other companies employees are rewarded for simply agreeing with whatever the boss says. Needless to say, opinions or judgments reached by the former sort of community are more credible than those reached by the latter. Equipped with these concepts, the non-expert can thus consider a candidate SC's judgments. A non-expert who considers the medical community's claim that cigarette smoking causes cancer may reflect first of all on the fact that the research scientists in that community are employed by colleges and universities that have no vested interest in the particular claims their scientists make – their primary vested interest is in the ability of such claims to survive severe scrutiny from other members of the community. This means that scientists have a deep interest in not endorsing claims that cannot be shown to be defensible from the standpoint of widely agreed upon principles of truth-conducive reasoning. That some scientists make such endorsements thus testifies to the existence of such principles. Scientists will be rewarded for making claims which will prove to be robust in the face of severe criticism, as well as for pointing out the ways in which the arguments of other scientists fail to respect the principles of truth-conducive reasoning. On the other hand, the non-expert who considers the claims of scientists funded by the tobacco institute is prepared to dismiss the conclusions of such scientists (either in the form of reports of projects or their consensus) on grounds of the non-expert's knowledge of the reward structure such scientists operate under.[8]

---

[8] Shapin (1994: 413) quotes Robert Merton to the effect that "the activities of science are subject to rigorous policing, to a degree perhaps unparalleled in any other field of activity." (1973: 275–276)

But while the consensus of an epistemic community can function as an important form of evidence for non-expert theory evaluation, it can be problematic as well. This is because the various conditions mentioned above that must be met to establish the probative weight of consensus are not always met, and it is not always easy for non-experts to determine whether they do or do not hold. There are various historical examples of SCs whose consensus was held in place by something other than the pure weight of the relevant evidence. One well-known travesty concerned the phenomenon of Lysenkoism in Soviet science (cf. Soyfer 1994). The genetic theories of T. D. Lysenko were never well supported by any actual evidence, but thanks to the political might of Joseph Stalin Lysenko's theories were taught in the Soviet Union as though they were established knowledge for several decades. Lysenko's theories predicted that Soviet agriculture could be reformed so as to substantially improve the crop yield of Soviet farmers – an improvement that was badly needed. The Soviet government's sponsorship of his career (which included demoting or firing any scientist who publicly denounced Lysenkoism) was apparently based in part on the fact that Lysenko told the authorities what they wanted to hear, and in part because Lysenko appeared to be the sort of 'common man who made good' that Communist propaganda loved to promote. Opponents of Lysenkoism were denounced as disloyal to the communist cause and used as scapegoats to help explain the failure of Stalin's program of agricultural collectivization. Under such conditions, of course, scientific consensus loses its probative weight. Consensus must be unforced, or it does not count. It is unclear, however, to what extent the non-scientific public of the USSR understood that the consensus surrounding the genetic theories of their scientists was the product of political force. It may have been that the general public assumed that scientific consensus was in general unforced – in which case non-expert theory evaluation based on such consensus obviously went awry.

One hopes that stories like that of Lysenkoism in science are rare, and it seems probable that consensus enforced by Stalinesque tactics is rare indeed. However, there remains the prospect of more subtle forms of coercion operating within the scientific community to produce consensus. One example concerns the forming of the Royal Society in seventeenth century England. Shapin and Schaffer portray the emergence of the Royal Society as partly the result of Robert Boyle's attempt to create a scientific community that would legitimate the rising experimental approach to natural philosophy (1985 Ch. 2). It was important for this purpose that there be a consensus among such a community that performing experiments established reliable knowledge of so-called 'matters of fact.' While the Royal Society came to be such a community, it is important to appreciate that acceptance of experimentalism was in effect a condition for membership in the Society! It was little wonder, then, that the Society reached consensus about the validity of the experimental approach to science. Under certain conditions, consensus can be subtly, or not so subtly, coerced. Consensus is an imperfect basis, though nonetheless in many cases an important one, for non-expert theory evaluation.

Consensus about the truth of some proposition (or a report of a particular project which claims such truth) offered by an SC that apparently meets conditions noted above thus seems to count, for non-experts, as a prima facie case for accepting such a proposition. But the prospect of the possible use of force in the generation of a community's claims should provoke a certain caution on the part of rational non-experts, who must remain alert to the possibility that scientists are responding to such force. But there is another form of evidence that is in some cases available to the non-expert and which can in principle alleviate the worry about the prospect of possible force, and it is novel confirmation. Let us consider again the case in which our SC collectively endorses T on the basis of some body of antecedently established evidence (or 'old' evidence) O. Let us now suppose that T entails some surprising observation statement N that was not appealed to by SC in its act of endorsing T and that the non-expert somehow knows this. N is now determined to be true, and thus constitutes an endorsement-novel confirmation of T. The question I want to pose is how the non-expert processes the epistemic significance of this event. Now it seems clear, given the facts, that N should confirm T for the non-expert – assuming of course that the non-expert has not previously been rendered absolutely certain of T. This follows simply because the non-expert knows that N is a true consequence of T and because the non-expert believes there is less reason to think that N will be true if T is false (say because he has been reliably informed that N is not otherwise known to be true and that it has been touted as a 'prediction' of T).

But is predictivism true for the non-expert? This is to inquire whether N, given its novel status, supports T more strongly for the non-expert than it would have had N been part of the body of original evidence O given to SC, or more strongly than a comparable piece of evidence which is actually part of O (assuming the non-expert were presented with O, or informed of its existence). My position is that the non-expert's view of this matter will depend in part on her view of the relevant SC. For example, one reason that is given on behalf of predictivism is that it is much 'harder' to endorse a theory that goes on to entail novel successes than it is to build a theory to accommodate a body of old evidence. As the old saying goes, "it is always possible" to build some theory to fit any given body of data – thus one should not be too impressed by evidence for a theory the theory was built to fit. This argument for predictivism could be based on a suspicion that theory builders will indulge in ad hoc theory endorsements – endorsements of theories which fit the known data but which are otherwise poorly supported by extra-empirical criteria or background belief (cf. 1.4.3). Now the question whether the non-expert will be a predictivist on a basis such as this has much to do with her view of the SC in question. For example, suppose she views the SC as a biased community who is probably willing to distort its judgments for the sake of self-interest. This could be because of political force of the types mentioned above, but the force could easily be economic as well. Suppose, for example, that SC is a community of scientists funded by the tobacco industry and T is the conjunction of the claim that smoking causes cancer together with various auxiliary claims. The non-expert may well suspect that this SC has found a way to accommodate the total data so as to preserve a commitment to the negation of T by adopting ad hoc auxiliary claims that serve to maintain ∼T while accommodating much data that appears to support ∼T (e.g., the auxiliary claims could include statements like "Data that apparently supports the carcinogenic quality of smoking were collected in misleading ways"). Thus (the non-expert reasons) ∼T is poorly supported because the theory building process very possibly incorporated ad hoc maneuvers – given the biases of the SC involved. However, suppose '∼T' went on to generate an authentically novel prediction N that was then confirmed. The sort of doubt that the non-expert has about T would not apply to this case, as the novelty of N precludes ∼T's having been built to fit it, or rather, ∼T's having been endorsed on its basis. Thus N would confirm T more strongly than the data the SC was initially presented with – so predictivism would hold true.

The previous paragraph shows how a certain sort of predictivism can be nurtured by a certain kind of non-expert skepticism – skepticism about the objectivity of the SC. Similar skepticism could be nurtured by the view that the SC has other kinds of vested interest in the truth of some theory T – if, e.g., a particular SC stands to gain prestige by the empirical vindication of T insofar as T was the creative product of that SC, or if that SC previously strongly endorsed T, etc. (the SC may have acquired what Popper called a 'dogmatic attitude,' cf. section 3.2). Such skepticism could also be generated by the non-expert's belief that SC is poorly qualified to investigate the issues they investigate – a lack of training in the relevant science may engender a willingness to accept theories that fit the data but are poorly supported by extra-empirical criteria, for a lack of training may consist of a lack of training in the use of such criteria. Successful novel confirmations are particularly important when they come out of such communities because they have at least some power to transcend this sort of skepticism. More precisely, such confirmations are important because they encourage the non-expert to revise her initial skepticism about the SC. This is the problem with accommodated evidence: it is less able to change the non-expert's preliminary assessment of the intellectual credibility of the SC. This is because there is a reasonable worry that any particular accommodation may be, in the phrase adopted in Chapter 1, a dubious accommodation.

Above I noted that consensus can be a fallible form of evidence for pluralist evaluators in part because consensus can be forced in various ways. One way of mitigating concern about the possibility of such force is to look for subsequent novel confirmations of the theories that the SC has endorsed. Lysenkoism in the USSR was one example of a forced consensus – but careful observers of Soviet agriculture who had been guided by Lysenko's proposed methods of seed vernalization (such as the farmers who were forced to use them) could not help but note that the many promises made by Lysenko about improved crop yields systematically failed (e.g., Soyfer 1994 Ch. 6). The absence of novel confirmation of Lysenko's theories thus should have suggested that the 'consensus' about them was misguided. On the other hand, Shapin and Schaefer argued that the consensus of the Royal Society about the soundness of the experimental method of science was at least somewhat forced, but the vast subsequent empirical successes of this method vindicated this consensus after all. The endorsement by an SC of some theory can thus serve as a presumptive argument in favor of the theory for a non-expert – subject to confirmation or revision by subsequent testing of novel consequences.

But consider now a different case: suppose the non-expert holds SC in the highest regard. SC is, as the non-expert sees it, a 'mature science' – one filled with highly educated and competent individuals who are ruthlessly objective in their pursuit of truth and committed to a highly developed and finely reticulated set of background beliefs (this field has reached what Psillos calls a 'take-off point' ((1999: 107–108), a concept he credits to Boyd). There is (almost certainly) no inappropriate force in this community that could produce a forced consensus. Notice now that the non-expert is less likely to suspect this SC of indulging in ad hoc theory accommodations – or of being so poorly educated as to not recognize ad hoc maneuvers when presented with them. Thus the non-expert seems less likely to accord novel confirmation any sort of special status – at least insofar as predictivism is motivated by the non-expert's suspicion of adhocery. Would the non-expert thus cease to be a predictivist? At this point let us recall that we are considering an SC that is characterized by a shared body of background belief K. It is K that guides the deliberations of the SC, and K which (together with O) led this SC to endorse T. Now we have assumed that the non-expert holds this SC in the highest regard – thus presumably the non-expert has at least a reasonably high degree of confidence in K, despite the fact that as a non-expert she does not apprehend the content of K herself. However, this is not to say that the non-expert regards K as certainly true. She may regard K as being reasonably probable, as she does not regard mature sciences as infallible in the choice of background belief. But now this SC has endorsed a theory T that has subsequently enjoyed novel success in the form of N. Now there is nonetheless a case for predictivism on the part of the non-expert: N now appears as a form of evidence that K is true. For it was K, along with the accommodated evidence O, that prompted the endorsement of T – new evidence for T thus verifies K. But the evidence O lacks this virtue from the non-expert's standpoint – for as far as the non-expert can tell, O may simply be evidence that renders T endorsable from the standpoint of someone who accepts K and provides no actual evidence for K itself. Of course it is possible that O does provide as much (or even more) support to K than does N – from the standpoint of the SC experts themselves (insofar as the experts are prone to consider the evidence for their own background beliefs, cf. section 2.4.4). O may also confirm T more strongly for the experts themselves than N. But the non-expert will probably lack sufficient knowledge to establish whether this is true – and thus should fall back on the principle of caution which assumes the weakest plausible evidential connection between O, T, and K. (The asymmetry between predicted and accommodated evidence with respect to background belief suggested here will be a central topic of Chapter 3.)

In the previous section I noted that consensus was, while an important form of evidence accessible to the non-expert, nonetheless a fallible one. The same is true of novel confirmation. This is because the appearance of novel success by some theory can constitute a facade that may deceive the non-expert. One classic example of such maneuvering is contained in the book *Chariots of the Gods?* by Erich von Daniken. Von Daniken argued in this book that intelligent aliens had visited the Earth during its pre-history. The primary form of evidence he presented consisted of numerous examples of archaeological or anthropological findings that seemed to many readers to be difficult to explain in other terms. These include the many myths of the Gods held by pre-historic people, and artifacts like the statues of Easter Island and temples of Peru whose construction by primitive peoples is admittedly astonishing. Giere explains the seductive power of von Daniken's presentation of evidence as due to its "building the hypothesis out of facts that are not known to the general public and that, therefore, are improbable given what most people know. One has the illusion that the hypothesis predicts improbable events. But . . . the events in question were known to the author and other experts all along, and were in fact used by the author in constructing his hypothesis." (1984: 160) The statues on Easter Island were not novel predictions of von Daniken's hypothesis, as they appear to be given the way the book is written, but were in fact accommodated by him in the process of constructing his theory.

Stories like the preceding point to an important moral about appeals to non-expert theory evaluation. The claims that constitute N must not simply be 'novel' from the standpoint of the evaluator – they must be novel from the standpoint of the endorser. This is to say that observation-based knowledge of N must not form part of the basis on which X endorsed T, and somehow the non-expert must establish this to be true. Admittedly, the evaluator's ability to determine that N is novel for X will be, in some cases, fallible. However, one should not worry too much about possibilities like these. The case cited above, for example, involves willful deception on the part of a pseudoscientist like von Daniken who hides important information so as to make himself appear a better predictor than he is. Willful deception on the part of individuals, while all too common, carries certain risks of exposure that many candidate experts will carefully avoid (so as to avoid, e.g., suffering von Daniken's fate of being publicly iden-tified as a crackpot). In other cases, the genuine novelty of an endorser's prediction can be essentially guaranteed because it is simply obvious to all concerned that the truth of the prediction was only established after the endorsement. These will be cases in which the prediction is novel in the temporal sense. This was presumably true, for example, of the first suc-cessful shuttle launch. Of course it is possible that prior to the first shuttle mission there were multiple secret and failed shuttle missions, so many that NASA finally hit upon a set of engineering methods that proved successful 'by accident' as it were (i.e. with no prior act of endorsement that that mission would be successful), and then followed with a publicly announced shuttle launch which proved successful and which simply mimicked the secret successful mission. In such a case NASA would have created a public facade of novel success. But such a hypothesis is pure paranoia. Similar remarks apply to other 'first successes' of scientific or engineering feats that were publicly displayed and preceded by clear endorsements that success would be achieved.

I have sketched a model of pluralist theory evaluation that focuses on the epistemic significance of consensus and novel confirmation. But there is an important ambiguity in this model that should be noted. On the one hand, a pluralist can be thought of as a person whose primary interest concerns what probability should be assigned to a particular theory T – and who takes acts of probability posting by competent individuals or communities to be epistemically significant in just this regard. However, as we have seen, the attempt to assess this epistemic significance will lead the evaluator to consider the background beliefs K of the individual (or community) that led him (or it) to post that particular probability. Thus the attempt to determine the probability of T leads the evaluator to consider the probability of K. As the posting of a probability can count as an epistemically significant act for the pluralist evaluator, so the possession of background beliefs can be epistemically significant as well: the pluralist maintains that the possession of such beliefs (on the part of well certified individuals) is a kind of evidence in favor of such beliefs. In some cases, as we will see, the pluralist evaluator is actually more interested in the probability of the background beliefs involved than she is in the probability of the theory in question. One common misunderstanding of the thesis of predictivism, in my view, is that predictivism is only important insofar as it illuminates the confirmation of individual theories. In fact, predictivism in some cases is important because of the critical role it plays in the assessment of background belief. To cite an example that will be taken up in Chapter 4, proponents of scientific realism like Richard Boyd argue that the instrumental reliability of science in endorsing theories that turn out to be empirically successful is best explained by imputing truth to the background beliefs of scientists which guide the scientific process of inquiry, deliberation, and theory testing. Likewise, as I will show in Chapter 3, some nineteenth-century scientists came to see the entire field of chemistry (viz., a large set of background beliefs constitutive of a broad field) as confirmed by the successful predictions of Mendeleev's Periodic Table of the Elements. Thus the pluralist model of theory evaluation can count as a tool for assessing both the probabilities of individual theories and the probabilities of sets of background beliefs – and of course these two assessments are importantly connected.

I have sketched arguments to the effect that non-experts can assess the consensus of epistemic communities of particular theories, but must consider the possibility that such consensus is a product of distorting political, economic or intellectual force. I have argued that the concern about the possibility of forced consensus can be mitigated by the adoption of a predictivist method of theory evaluation based on different attitudes toward the relevant SCs. I will return to these issues in the next chapter, where I offer a rigorous formulation of them.

I hold that non-expert theory evaluation is as important to the world in general as expert theory evaluation. Non-expert theory evaluation is mandatory in an age in which individuals need access to knowledge that is possessed primarily by highly trained experts. But the importance of non-expert theory evaluation is even greater than these remarks suggest. We live in an age, as it is often said, which is dominated by science. This 'domination' consists of more than the widespread developments of technology that have transformed our lives in so many ways. Science is, in the minds of many, the ultimate and most prestigious source of knowledge in the world. Many people who are not themselves scientists have been transformed in their overall view of the world by a growing confidence, particularly among generally educated people, in the authority of science. Consider, for example, the impact of the growing acceptance of evolutionary theory by the scientific community in the latter part of the nineteenth century on the attitudes of non-scientists toward the authority of religious texts such as the Bible. While the conversion of the scientific community to some form of evolutionary theory took place quite quickly, the realization among the general public that this conversion had taken place took some time. The event which brought this realization home, particularly to the American people, was the Scopes trial of 1925. Particularly astonishing to the many observers of the trial was the inability of prosecuting attorney William Jennings Bryan to identify reputable scientists who would admit to believing in the Biblical account of creation. He presented only two, and their qualifications were sketchy. The realization that the scientific community at large endorsed evolutionary theory and repudiated the literal Biblical account of creation was a devastating blow to many proponents of Biblical fundamentalism. One important reaction to this event was the growing attempt on the part of mainstream Christian denominations to make room for the truth of evolutionary theory by foregoing a literal reading of the Book of Genesis (Marsden 1980: 184–195). But there is no reason to think that all or even most of the theologians, pastors and members of such denominations themselves had any detailed understanding of evolutionary theory or the evidence that apparently supported it. They were non-experts. They deferred, ultimately, to the consensus of the scientific community. But such deference can be plausibly construed as an act of theory evaluation: the theory of evolution was judged by them to be probably true – or at least very possibly true – on the basis of a consensus of scientific opinion generated by a scientific community. One can scarcely overestimate the effect of a general powerful argument against Biblical veracity on the mindset of religious communities – or indeed nations – that have historically taken the Bible as the inerrant word of God.

Scientific research is, these days, a highly expensive activity that depends heavily on public and private funding. But insofar as the availability of funding for scientific projects rests upon the willingness of public officials and philanthropists to provide such funding, it depends upon the convictions of people that science deserves such funding – and such people are often themselves non-experts. Of course, bureaucrats or philanthropists who provide such funding may defer to the assessment of prospective research projects by experts – but their deference is just another form that trust (however provisional) in the overall accuracy of the theories such experts endorse takes. It is obvious that the bulk of ongoing scientific research – as well as the influence of science in general on people's metaphysical and religious views – depends in a profound way on the authority of science among non-experts. I conclude that non-expert theory evaluation is as important a phenomenon as expert theory evaluation to the sustained activity of science.

So far I have argued that non-expert theory evaluation is both important and at least prima facie rational – insofar as non-experts can appeal to a pluralist model of theory evaluation. In the absence of some model of non-expert theory evaluation, moreover, science is powerless to win the reasoned support of the non-expert population. But it is not my position that only non-experts can make use of this model. Experts too, of various types, find the pluralist model useful and in many cases essential. To this topic we now turn.

2.4 EXPERT PLURALISM

2.4.1 *Interdisciplinary experts and team-member experts*

Some scientific fields today are interdisciplinary fields – such fields are composed of experts who share an interest in certain issues and questions but possess disparate educational backgrounds, and thus different background beliefs (such as environmental science (Manahan 1997: 3)). Interdisciplinary experts thus are compelled to make important use of the pluralist model of theory evaluation in working with their colleagues.

But the phenomenon of non-overlapping background belief among colleagues working together to answer a single set of questions is not limited to fields usually thought of as interdisciplinary. It is increasingly true of fields like physics in which work is carried out by teams of experts who possess disparate bodies of expertise within their discipline. Hardwig cites an example of a paper that published the results of a physics experiment carried out by William Bugg together with 98 co-authors. Although all the co-authors were experimental physicists, they brought different specializations to bear on the project, and consequently "no one person could have done this experiment ... and many of the authors of an article like this will not even know how a particular number in the article was arrived at." (Hardwig 1985: 347) Such teamwork is even practiced occasionally in mathematics as well (Hardwig 1991: 695–696). Hardwig puts the point forcefully: "Modern knowers cannot be independent and self-reliant, not even in their own fields of specialization. In most disciplines, those who do not trust cannot know; those who do not trust cannot have the best evidence for their beliefs" (1991: 693–694) Epistemic pluralism is unavoidable for the interdisciplinary or team member expert.

2.4.2 *Imperfect experts*

It is well known that scientific expertise today is a highly specialized affair. For example, a PhD-bearing biologist will typically have a fairly tightly circumscribed sub-specialty, such as genetics, microbiology or evolutionary theory. But it will typically be the case that a scientist who qualifies as an expert in some sub-specialty F will not be familiar with all there is to know in F. The typical geneticist, e.g., will be familiar with a large body of background belief required for research in this field, and familiar as well with many theories, accepted and hypothetical, and the associated evidence that has been offered for them. But it may well be the case for the fully credentialed geneticist that there are theories or empirical data within his own subspecialty with which he is only partly familiar, or even entirely unfamiliar. These would include theories that are not regarded as particularly fundamental within his field, or theories which are highly innovative and not yet part of accepted belief, theories which for one reason or another the sub-specialist has not taken the time to study and evaluate. I refer to experts who lack perfect knowledge of their own sub-specialty as 'imperfect experts' – and I hasten to add my suspicion that the vast majority of sub-specialists are imperfect in this sense.

The pluralist model is fundamental to the deliberations of imperfect experts. In particular, imperfect experts may base decisions to study and come to understand new theories in their sub-specialty on the basis of an evaluation of a theory they do not as yet understand. This evaluation could focus on a consensus among certain experts in her own field together with announced novel confirmations of the theory itself – the imperfect expert thus regards the subset of colleagues who endorse the theory as an SC to which she does not belong, and regards their consensus as an important form of evidence for the theory. Subsequent novel successes confirm the judgment of this SC in basically the way described above: novel successes confirm that the experts who endorse the theory are not misguided. She thus evaluates the theory – initially – as a non-expert. Indeed some form of non-expert evaluation is essential to experts who must make intelligent decisions about which theories to invest time and energy in apprehending. They must then evaluate theories prior to understanding them – and this is what the pluralist model allows them to do.

Earman and Glymour's (1980) article provides a straightforward example of the use of the pluralist model by imperfect experts. This paper concerns the historical process by which Einstein's theory of relativity was evaluated and eventually accepted by the British physics community. In this case the British physicists play the role of the imperfect experts, while the relevant epistemic community consists of a small community of physicists including de Sitter, but especially Eddington and Dyson. Earman and Glymour explain that prior to 1919 British physicists were aware of the existence of Einstein's general theory, and aware that the theory had some proponents in the form of a small community consisting of Eddington and de Sitter. But for various reasons the British physicists did not take the theory particularly seriously, not even to the point of making a serious effort to understand what the theory said. Partly this had to do with the radically new content of relativity theory itself – though considerations of nationalism also played an interesting role. Hostility to Germany was, at the end of World War I, so intense that some British physicists had called for a boycott of German science. A strong reversal of attitude took place in 1919, however, thanks to the British expedition (headed by Eddington) that confirmed Einstein's prediction of starlight bending. Karl Popper, as everyone knows, cited this confirmation of relativity theory as a textbook example of a novel confirmation (Popper 1963: 34). The fact that the successful expedition was British, Earman and

Glymour claim, helped British physicists overcome their nationalist antipathy toward relativity theory. But two other factors were of crucial importance: the endorsement of the theory by a small but distinguished sub-community of physicists, and the knowledge that the theory had enjoyed an important novel confirmation. In this case these two factors were combined – the theory had early proponents in de Sitter and Eddington. However, after the eclipse expedition of 1919, it was Dyson and Eddington who persuaded the British physicists that the data obtained by the expedition actually supported Einstein's theory (Earman and Glymour 1980: 50). (This claim, as it turns out, was by no means obviously true (1980: 76–81)). Eddington much later claimed that "the announcement of the results [of the eclipse expedition] aroused immense public interest, and the theory of relativity which had been for some years the preserve of a few specialists suddenly leapt into fame." (1980: 84)

There is every reason to believe that this sort of use of the pluralist model by the British physicists is typical of all scientific fields. Experts typically do not have the time, energy or interest to study every emerging theory in their field – hence most experts are imperfect experts. Imperfect experts have an important resource in the pluralist model of theory evaluation.

2.4.3 Humble experts[9]

Imagine an expert who possesses the relevant background beliefs K required for competence in her particular sub-specialty, and who considers what probability should be assigned to some theory T. Unlike the inter-disciplinary expert and the imperfect expert, let us suppose there is no further body of knowledge she needs to acquire in order to evaluate the theory herself. Suppose that after careful consideration she reaches a conclusion about what probability should be assigned to T. Assuming that she believes herself possessed of all background beliefs that are available that bear on the probability of T, and assuming she computes T's probability carefully, she might come to regard the posted probabilities of other agents as epistemically irrelevant. However, a scientist who actually possessed all the relevant information might nonetheless worry that her posted

---

[9] White (2003: 672–673) describes a kind of theory evaluation which resembles my picture of the evaluation performed by the humble evaluator – one concerned about the possibilities of computational error and the inaccessibility of endorser reasoning – and suggests that this kind of evaluation could play an important role in the truth of predictivism.

probability for T is erroneous because some critical bit of information might have been forgotten. Consequently she will be assuaged if other competent scientists post probabilities that agree with hers, or concerned if they do not. Because the possibility of such forgetfulness is rather high when there is a large body of relevant background information at stake, an expert may adopt a pluralist approach out of a certain spirit of epistemic humility – I call such experts 'humble experts'. Another source of epistemic humility is in the humble expert's recognition of her susceptibility to computational error. Suppose that the determination of T's probability involves a large number of mathematically challenging steps – in which the possibility of error even for careful calculators is reasonably high. Thomas Reid illustrated this point with the following thought experiment:

Suppose a mathematician has made a discovery in that science, which he thinks important; that he has put his demonstration in just order; and, after examining it with an attentive eye, has found no flaw in it . . . He commits his demonstration to the examination of a mathematical friend, whom he esteems a competent judge, and waits with impatience the issue of his judgment. Here I would ask again, Whether the verdict of his friend, according as it has been favorable or unfavorable, will not greatly increase or diminish his confidence in his own judgment? Most certain it will and it ought. (From Thomas Reid, *Essays on the Intellectual Power of Man*, in *Thomas Reid's Inquiry and Essays*, ed. R. Beanblosson and K. Lehrer, 6, 4, 262–263, quoted in Foley 2001: 115)

Epistemic contexts in which the estimation of probabilities is confounded by the sheer size and complexity of the set of relevant background beliefs or by the difficulty of the relevant computations are contexts in which theory evaluators concede their own fallibility and consequently depend more heavily on the judgments of other competent agents. It seems quite reasonable to suppose that much theory evaluation is carried out in such contexts, and thus that much theory evaluation is performed by pluralists.

It may be objected that the pluralism of humble experts will be momentary – if another agent posts a probability that is significantly different from that of the humble expert, this will simply direct the humble expert's attention to whatever reasons underlie the judgment of the other agent. But my position is that there will be multiple cases in which the reflections of the dissenter may resist complete articulation, as they may include modes of analysis or particular insights that have been acquired over years of scientific practice, in which case expert pluralism will prove a more enduring – possibly permanent – tendency in a suitably humble evaluator. Walton suggests a similar point in the form of what he calls the 'inaccessibility thesis' which claims that expert judgments cannot be traced back to some set of premises and inference rules that yield the basis for an expert judgment. He suggests "that expertise tends to be a somewhat elusive quality. An expert may not understand why she does things in a certain way, but may only have an instinctive understanding that she is doing it in the right way. Hence her ability to explain why she does things in a certain way may be limited." (Walton 1997: 112) Such ineffability may well prevent an expert from directly inspecting and considering another expert's epistemic basis for a judgment based on a complex body of information. Speelman (1998) notes that there is a considerable body of empirical data that supports the inaccessibility thesis and points to systematic reasons why expert knowledge often becomes inaccessible to verbalization. These include the fact that individual steps in expert reasoning may never enter the expert's awareness or are too fast to be noticed, or because expert knowledge is stored in spatial, pictorial or imagelike forms which resist explicit description.

2.4.4 Reflective experts

What qualifies an individual as a member of some SC is primarily the possession of a particular body of background belief K. It is the purpose of a scientific education to instill K in developing experts. But what is the attitude of the expert himself toward K, which presumably contains his own beliefs? Insofar as they are actually his beliefs, he presumably regards them as true (or perhaps just empirically adequate) or as sufficiently probable to be regarded, however tentatively, as true. Furthermore, as a working expert, it is not typically his purpose to evaluate the various beliefs in K – rather, he assumes them, often implicitly, in the course of investigating detailed problems on the edge of scientific research. This is true partly in virtue of what we mean by 'background beliefs' – the term suggests beliefs that are not the focus of deliberation or test, but which serve to guide scientific experimentation, discovery and theory evaluation. It is true too because of the nature of scientific education, which typically presents bodies of widely accepted theory in textbook form, and demands mastery of the given material while discouraging critical reflection about whether the evidential support of the material is sufficient to warrant belief. There is good reason for such discouragement, for success as a scientist typically results from a deployment of accepted background belief in the course of investigating new problems, not by way of reflection on fundamentals (cf. Kuhn 1970 Ch. 4). In any case, that a scientist is not acquainted with the evidential ground of all his basic beliefs is practically guaranteed by the limitations on his time and energy, and by the threat of infinite regress (for every new proposition he learns, he must learn *its* evidential basis, and so on). To learn the evidential basis for many accepted basic theories in his field of expertise would require scouring historical texts, and that is time that could be spent solving new problems.

But what then should we say about the expert's belief in K? Is it simply unjustified, given his at least partial ignorance of its evidential basis? A scientific expert might indeed simply never raise the question whether his own background beliefs are justified, and pass through his career quite successfully nonetheless. But some experts might inquire what actual basis they have for their various background beliefs, given their inability to know the details of the epistemic basis of all or even most of them. Such inquiry might be induced by pressure from science skeptics (e.g. religious fundamentalists or radical post-modernists) as to what reason he has to believe in the overall credibility of the scientific enterprise as he practices it. I refer to such experts as 'reflective experts.' Can reflective experts, by some approach, regard themselves as justified in believing their own K's?

The most plausible approach, I claim, is for the reflective expert to adopt the perspective of a pluralist. As a pluralist, he can appeal to certain facts about the scientific community that has collectively endorsed K that are epistemically accessible to him. This would include, first of all, the fact that the various theories and beliefs that constitute K are claims that have enjoyed consensus among the various expert evaluators who evaluated these beliefs directly, viz., on the basis of the evidence for them. Consensus means to the expert essentially what it means to the non-expert – it indicates the presence of powerful evidence on the part of the participants in the consensus. The expert can possess evidence that the community that produced this consensus is capable, well educated, unbiased and unforced in their theory evaluations – even more so than the non-expert, given the expert's greater familiarity with expert reasoning. Likewise, the expert can appeal to whatever record of novel success the relevant SC has enjoyed – any, of course, with which he is aware – throughout its history. Such novel successes could include successful predictions of particular phenomena made by his predecessors in the field which are brought to his attention, but could include as well the novel successes of his own theory endorsements, insofar as his theory endorsements are guided by his commitment to K, and thus his novel successes reflect as well as any other on K (we will revisit this point in our study of Mendeleev in the next chapter). My point is that the logic of K evaluation by the reflective expert is essentially that of the pluralist model – for this is, in practical terms, the best he can do.

This is not to deny either that there are cases when the epistemic ground of widely accepted background beliefs does take center stage in some experts' reflections – these are the occasional episodes in which established background beliefs are on the verge of giving way under the pressure of accumulated anomalies in favor of a very different set of background beliefs, the sort of event Kuhn called a scientific revolution. But these episodes are, it is generally claimed, few and far between. And even experts who figure crucially in revolutions will find themselves unable, for the very reasons mentioned above, to evaluate exhaustively the evidence that supports whatever background beliefs they maintain throughout the crisis period. There is simply no alternative for reflective experts than to adopt the pluralist's approach to theory evaluation vis à vis all or at least some of his own background beliefs. Insofar as it is the pluralist model of theory evaluation that justifies the reflective expert's belief in K, it serves as an epistemic foundation for his scientific beliefs.

Thus the interest of the pluralist model is not limited to its applicability to non-experts who need, for one reason or another, to evaluate certain scientific beliefs. The model has much to offer interdisciplinary, imperfect, humble and reflective experts as well.

2.5 CONCLUSION

I have argued in this chapter that much scientific theory evaluation is pluralistic. This is obviously true at the level of non-expert theory evaluation. I have argued, moreover, that pluralist theory evaluation is important for expert scientists themselves. But even this is not an entirely novel claim – for there was Kuhn's old adage that there is, with respect to the assessment of a theory, no higher court of appeal than the assent of the relevant community clearly. The adage sounds a pluralist tone, and will be true not only for evaluators outside the community but for experts inside it as well: the highest form of re-assurance that an expert's posted probability is correct is that her community agrees with her. If she assigns such significance to the community as a whole, she must assign at least some significance to the judgments of individual members of the community (else the community would be a community of incompetents and could not have the authority she imputes to it). Expert pluralism, particularly in epistemically complex contexts, is surely not an unscientific attitude.

I have argued that expert pluralism pervades ordinary scientific practice. But this phenomenon is particularly visible when we consider the impact of a particularly prestigious scientist who publicly endorses a particular position. Sobel (1987: 8) argues that the particle theory of light held sway among English physicists during the eighteenth century despite the existence of powerful arguments for the wave theory (such as those of Christian Huygens) due to the eminence of Newton, who had endorsed the particle theory. Likewise Barrett (1996) argues that the history of quantum mechanics was critically affected by the public endorsement of particularly prominent scientists. And there is no question that authors of scientific journals do appeal to the judgments of prominent scientists in making their cases (cf., e.g., Achinstein 1991: 80), suggesting that authors conceive of their readers as pluralist evaluators, at least to some degree. Steven Shapin's *A Social History of Truth* provides a detailed and scholarly demonstration of the reality of expert pluralism in seventeenth century science, and he concludes that expert pluralism is no less fundamental to contemporary science (despite some important differences in the form such pluralism takes). He concludes that "Scientists know so much about the natural world by knowing so much about whom they can trust." (1994: 417)

I attribute a modest pluralism to experts – experts do typically make their own assessments of theory probabilities based on their best assessment of the total evidence, but also take the posted probabilities of other experts whose opinions they respect to have epistemic significance. It is an interesting and complex question how scientists combine the content of their individual probabilities with those of other experts. This issue will be explored in Chapters 3 and 6.

I claimed in Chapter 1 that a pluralist approach underwrites a tempered predictivism, and suggested further in this chapter, in somewhat broad terms, why this is so. We turn in the next chapter to a rigorous exposition of this claim, and a detailed historical illustration of it.

CHAPTER 3

Predictivism and the Periodic Table of the Elements

3.1 INTRODUCTION

Recall the happy predicament you imagined yourself to be in at the beginning of this book: you are, after years of subsisting on rice, beans, and basic cable TV, suddenly wealthy. You search for a financial advisor to help you invest your money, and choose an advisor with a strong record of predictive success over another who explains the same market fluctuations only ex post facto. This seems like an eminently wise choice, but why? Somehow or other, predictive success serves as an indicator that predictors know what they are talking about. It is time to get more precise about this claim.

In the previous chapter I sketched a pluralist model of theory evaluation and argued that pluralist theory evaluation is both widespread and important to the scientific community in a variety of ways. One reason for developing this model was to help undermine what Hardwig calls the romantic ideal of epistemic individualism, a myth I suspect still has some currency among philosophers of science. But the other reason for developing this model was to pave the way for a new account of tempered predictivism, one tailored to pluralist theory evaluators. The basic logic of this new account is suggested in the previous chapter: successful predictions, unlike accommodations, can serve two distinct purposes for the pluralist evaluator: they escape suspicion of adhocery and can provide a special kind of confirmation of the background beliefs of endorsers. In this chapter I give a more detailed rendering of the pluralist-based theory of predictivism. In what follows I distinguish between two types of endorsers that I refer to as 'virtuous' and 'unvirtuous' – and provide a distinct account of predictivism for each type. I then turn to a detailed illustration of my two accounts by focusing on the well-known case of the emergence of Mendeleev's Periodic Table of the Elements, a story that has been cited by various philosophers as a classic illustration of the reality of predictivism among working scientists. I then compare my account of this story with other accounts based on other theories of predictivism.

But before proceeding I should emphasize that it is not my intention to argue only for tempered predictivism. It is one of my theses that a typical community of evaluators will be a diverse lot. Some theory evaluators exhibit thin predictivism in their assessment of theories – others tempered predictivism. I will provide a detailed account of both tempered and thin predictivism in this chapter – both accounts are, as noted in (1.5), versions of weak predictivism. I repudiate strong predictivism – a position I will defend in detail in the next chapter.

3.2 UNVIRTUOUS PREDICTIVISM

Let us begin with tempered predictivism. I will consider a pluralist evaluator, Eva, who knows that some observation statement O is true and notes that theory T, which we assume was constructed by a random theory generating machine, entails O – T also entails another heretofore untested and unestablished observation statement N. Consider now two distinct scenarios: in the first, Eva encounters another agent, Peter the predictor, who she believes to possess a certain degree of competence as an appraiser of theories, who posts an endorsement level probability for T on some evidential basis that Eva knows includes O – he thus predicts N (as yet unestablished). Thereafter N is shown true. In the second scenario, Alex the accommodator – who is regarded by Eva as precisely as competent as she regards Peter in the other scenario – refuses to endorse T on the evidence Alex possesses (which Eva knows includes O) – but does post an endorsement level probability for T after N is observed to be true. How should Eva's assessment of T's probability compare between the two scenarios? Clearly it will matter to Eva not only on what basis the theorists post endorsement-level probabilities, but which particular probabilities they post. Since she regards both scientists as competent, and since she is a pluralist, her probability for T should be higher the higher the probability either scientist posts. Suppose, for the moment, that in both scenarios, once N is shown true, the two scientists declare that the theory has a particular probability Q. Might she accord T an enhanced probability in the first scenario – despite the fact that she regards the two scientists as equally competent, and given that they ultimately post one and the same probability on the basis of the same evidence (as far as Eva can see)? One suggestion that was noted in Chapter 2 is that there is a widespread tendency on the part of evaluators to view accommodators with a peculiar form of suspicion. For it is tempting to think of the accommodator as though he is something like a Popperian pseudoscientist or a perpetuator of a Lakatosian degenerating problem shift – a person, that is, who holds on to a failing theory by repeatedly performing ad hoc repairs that accommodate recalcitrant data. As noted already, one of Popper's classic examples of this sort of pseudoscientist was the neo-Marxist of Popper's day – who would willingly fix Marx's theory any time its predictions were falsified by contemporary events to produce a new version of Marx's theory. Such pseudoscientists, Popper claimed, possessed a 'dogmatic attitude': they clung tightly onto their beliefs and tried to verify them at every turn. This dogmatic attitude apparently arises on the part of someone who is overly committed to a particular theory. This could be because his reputation is tied to the success of that theory – either because he is the originator of the theory or because he publicly endorsed it. This kind of dogmatic attitude on the part of an accommodator could cause the accommodator to do one of two things when he posts a probability for the theory on the basis of some evidence E: he could deliberately misrepresent what probability he actually thinks is justified, given E, and publicly post a probability for a theory higher than what he sincerely accepts, or he could even fool himself into thinking that the probability of the theory on E is higher than it actually is, given his total belief and evidence. Thus the accommodator would be guilty, in posting his probability, either of logical dishonesty or logically fallacious reasoning. I will call posters of probabilities who are either logically dishonest or inept 'unvirtuous' – the virtuous poster of probabilities is both logically honest and logically omniscient (in the same way as the Greek term for virtue, arête, suggests a mixture of morality and competence). A virtuous poster of a probability for some theory, in other words, posts the probability for that theory that accurately reflects the evidence he has that pertains to that theory (I give a more precise explication of the 'virtuous poster' below).

Another way of rendering the distinction between the virtuous and unvirtuous poster is in terms of the distinction between coherent and incoherent beliefs. The virtuous poster posts a probability that coheres with his other doxastic commitments – the unvirtuous poster does not, for he is guilty of a logical misstep. This does not imply that a virtuous endorser holds a set of beliefs that is entirely coherent – only that his posted probability for T coheres with all beliefs he holds which are relevant to T. This entails that the beliefs a virtuous endorser holds that bear on T's probability must themselves be coherent. If they are not, then no posted probability for T coheres with the relevant beliefs, and any such endorser who posts a probability for T must be considered unvirtuous.

Predictivism could now be defended as a strategy by which a pluralist evaluator protects herself against unvirtuous accommodators. For while there is, under certain circumstances, a tendency to worry that the N-accommodator may engage in unvirtuous behavior when he endorses the theory, there is less corresponding fear about the N-predictor. Since the N-predictor, when he endorses T originally, does not make use of any knowledge of N, his endorsement of an N-entailing theory cannot have been driven by any dogmatic desire to hold onto some earlier version of the theory that he modifies (in some suspicious way) to accommodate N, or by his desire to simply endorse any N-entailing theory. Thus there is less reason to think that his initial endorsement level probability for T was inflated by irrational factors – such as dishonest posting or wishful logic. An endorsement coming from an N-predicting T-endorser thus carries more weight than an endorsement from an N-accommodating T-endorser – so predictivism holds true for the pluralist evaluator. This theory of why predictivism holds is similar to what Lipton calls the 'fudging explanation' of predictivism (Lipton 2004 Ch. 10). On this basis, Eva could accord a higher probability to T in the Peter scenario than in the Alex scenario – say because she regards both Peter and Alex as under pressure to endorse an N-entailing theory, but only Alex could respond to such pressure by proposing an ad hoc theory rescue or ad hoc hypothesis (cf. 1.2.1, 1.4.3).

While this sort of reasoning has some plausibility, as a general account of pluralist theory appraisal it is unfair – for it amounts to anti-accommodator bigotry. Surely not all N-accommodators will be unvirtuous. It is perfectly possible to imagine an agent who posts an endorsement level probability for T on the basis of N who is both sincere and logically competent – say because he is under no social pressure to endorse some N-entailing theory or to show that his pet theory can accommodate N. It is thus important to consider the question whether predictivism can hold true for a pluralist evaluator who regards endorsers as virtuous – as will be shown in the next section, this is certainly possible.

But before proceeding we must consider the corresponding case of the individualist evaluator – who confronts Alex and Peter as before, but who assigns no epistemic significance to their acts of theory endorsement. Suppose, e.g., that the evaluator takes herself to be fully informed of the relevant evidence, including any evidence that is possessed by Alex and Peter and which prompts them to post their probabilities. Peter posts an endorsement level probability for T on some basis that does not include having observed N, as before, but suppose Alex simply endorses T primarily on the basis of N – T is otherwise poorly supported given what Alex knows and does not actually merit endorsement. But we will suppose Alex is unvirtuous – he posts a probability for T that is endorsement level (perhaps due to the sort of social pressure noted above) though this represents a logical misstep: he should not endorse T given the evidence he has.

Now our individualist evaluator will formulate her own judgment about the probability of T in each scenario. But it may be that in this case there is nonetheless a thin predictivist effect. The individualist will discern a thin predictivist effect if she discerns a link between prediction and some feature F that enhances theory probability. The enhanced probability of T in the predictor's scenario is attributable to a connection between prediction and the existence of evidence – independent of N – that supports T. In the Peter scenario T is supported independently of N (by the background beliefs that prompted Peter's endorsement of T), whereas in the Alex scenario T is not sufficiently supported (i.e. it does not justify assigning a probability as high as Q) (cf. the conception of adhocery described in 1.4.3).[1]

Thus epistemic contexts in which evaluators view accommodators as at least somewhat likely to be unvirtuous and as subject to various social pressures are contexts in which a certain form of predictivism can hold – I will refer to this species of predictivism as 'unvirtuous predictivism' and refer to evaluators who demonstrate a predictivist effect for such reasons as 'unvirtue-ascribing predictivists.' Such predictivism, as we have seen, can take both tempered and thin forms. But at this point we must return to the question of predictivism in the other sort of case, one in which evaluators believe the endorsers to be virtuous. As we know, predictivism can hold in these contexts too, but for quite different reasons.

3.3 VIRTUOUS PREDICTIVISM

In what follows I will assume that all predictors and accommodators are virtuous in my sense: they post no probabilities for theories that are unjustified given the evidence they hold – they neither misrepresent their

---

[1] This is not to say that, in the Alex scenario, there is no evidence independent of N that supports T – only that there is too little evidence to merit the endorsement level probability that Alex posts once he is presented with N. This is just to say that Alex accommodates N by way of an ad hoc rescue or ad hoc hypothesis.

assessment of this probability nor commit logical errors in computing what this probability is. I begin again with the assumption of a pluralist evaluator – who furthermore assumes the endorsers to be virtuous (we assume the relevant context is one in which the various social pressures that produce unvirtuous posting are absent). I am of course supposing that, in such cases as I consider, there is some fact of the matter what probability for a theory is justified given the evidence possessed by a virtuous poster. I clarify and defend this assumption below. Let us imagine a two dimensional coordinate graph which diagrams the relationship between two observable variables. We now consider two scenarios involving Eva the pluralist evaluator. Once more, in scenario 1, Eva notes that a number of observations O have been made which are depicted on the graph in the form of data points – a random theory generating machine has constructed an irregular line T which fits O. Eva presents O and T to Peter the predictor, a competent appraiser of theories in her judgment, and inquires how probable he thinks T is based on O (and whatever other background belief he possesses, cf. below). Hereafter we will refer to probabilities posted by agents for T on the basis of O (and the agent's background belief, but not observational evidence for N) as the agent's 'prior probability for T.' Peter posts a prior probability P_1 for T that is above the contextually fixed limit L for endorsement (cf. (2.2)) (and above Eva's own assessed value of T, if she has one, as Eva is not herself an endorser of T at this point). Now a number of new observations are performed, producing a new body of data points N that fall neatly on T. Hereafter we will refer to probabilities posted by agents for T on the basis of both O and N (and the agent's background belief) as the agent's 'posterior probability for T.' Peter now announces his posterior probability P_2 for T – which is higher than the prior. At this point Eva judges that T should be accorded a posterior probability V_p in the predictor's scenario. In the second scenario, Eva presents O and T to Alex the accommodator, whose competence in her view is equal to Peter's (though Peter himself plays no role in this scenario), and inquires how probable Alex thinks T is at this point – Alex posts a prior probability A_1 for T that is below the limit L and thus does not endorse T. Now Eva presents to Alex the additional data N – and now Alex responds by posting a posterior probability A_2 for T that is above L. Eva judged that T should be accorded probability V_a in this scenario (the accommodator's scenario).

Suppose you were Eva. Should V_p and V_a have the same value – or is there a predictivist effect (i.e. should V_p be greater than V_a)? One could object at this point that the example as given is underdescribed – for we are not told precisely what the values of the various probabilities (P1, P2, A1 and A2) are, and this is obviously important information. For the moment I want to set aside the question of these precise values with the promise to return to them later – at this point I want to think of the example simply in terms of the basic notion of endorsement. In such crude terms, there is nonetheless a palpable intuition that Vp should be greater than Va, despite the fact that in the two scenarios a particular theory is presented and supported by the same body of evidence (as far as Eva can tell) and endorsed by theorists of equal competence – and despite the presumed absence of adhocery. But what underwrites the predictivist intuition in this example?

My position is that the intuition that predictivism holds true in such an example has two roots. To make this point we must focus initially on the prediction scenario and break it down into two components. One component I deem the 'prediction per se' – in the prediction scenario this is constituted by Peter's posting an endorsement-level probability for T without appealing to any observations that directly support N. Peter thereby predicts N. The other component I deem 'predictive success' – this is constituted by the observation of N's truth that constitutes the success of Peter's prediction. In the scenario as described the prediction per se precedes its predictive success, but this is not essential – N could have been determined prior to Peter's endorsement of T though Peter nonetheless predicted N by posting an endorsement-level probability for T without appeal to N. T would have been novelly successful from the very moment of its endorsement by Peter. With this distinction in mind one can motivate the intuition that predictivism is true in two quite different ways, depending on whether one focuses on the prediction per se or on predictive success. Let us begin with the prediction per se. Eva may reason roughly as follows: the very fact that Peter endorsed T without appeal to N suggests that Peter possesses reason(s) R for believing that T may well be true. Now of course we and Eva know that Peter possessed knowledge of O prior to endorsing T – but it is important to the example as described that it is not the case, for Eva, that O alone is sufficient to make T endorsable (recall Eva is not herself a predictor of N, but simply an evaluator of T). Peter's reason(s) R could consist of his knowledge of still more observations (other than those in O and N), or of the fact that T coheres with other accepted theory, or that T meets certain extra-empirical criteria for theory choice that Peter may be aware of, or of background belief that establishes that O is in fact very strong evidence for T, or some combination of these. Because competent scientists do not endorse theories on the basis of insufficient evidence (Howson (1988: 382)), Eva may judge that reason(s) R exists – though she either does not know what R is or, if she does, does not know what evidential basis supports R for Peter, or if she does, she does not know what evidential basis supports this evidential basis, etc. (we will designate whatever body of reasons are unpossessed by Eva as 'R').[2] Eva can now judge that T is supported not only by O and N, but also by R as well. If Eva is prepared to trust Peter as a provider of reasons at all, this suggests that predictivism holds true. The higher Peter's prior probability, of course, the stronger his act of endorsement, and thus the stronger his confidence in the predicted consequence N – and thus the greater the predictivist effect for Eva. This is because in the accommodation scenario, insofar as Alex is unwilling to endorse T on the basis of O, without further knowledge of N, Alex apparently possesses less in the way of independent reasons for accepting T (though of course he may yet possess some reasons that favor T is not possessed by Eva). But this means that T is less well supported for Eva in the accommodation scenario than in the prediction scenario, and predictivism holds. This version of the predictivist thesis I refer to as 'per se predictivism' – because what explains the difference between Eva's probability in the two scenarios is what she takes to be the epistemic import simply of Peter's prediction per se – his prediction testifies to the existence of independent reasons for accepting the theory. The stronger his predictive endorsement and the greater Eva's trust in Peter's competence, the greater the predictivist effect.

Let us now focus on predictive success. Peter has, Eva assumes, endorsed T on the basis of some reason(s) R independent of O and N. But suppose Eva, despite her respect for Peter's judgment, is not certain that R is true. The fact that Peter's prediction of N is successful may serve to confirm the truth of R for Eva. This is basically because the predictive success of a theory that is recommended primarily by R is more likely if R is true than if it is not. Alex the accommodator, by contrast, has not enjoyed any predictive success, and the absence of any post-endorsement predictive success leaves any doubts Eva may have about the truth (or existence) of Alex's reasons (other than O and N) unreduced. This version of predictivism I deem 'success predictivism' because what accounts for the predictivist effect is the epistemic import of the success of the predictor's predictions rather than simply the prediction per se. The greater the enhancement of the evaluator's confidence in the truth of the predictor's reasons, the greater will be the predictivist effect.

---

[2] Here we apply the basic point of the Keynesian dissolution of the paradox of predictivism (cf. 1.4.3).

The story thus far has been described from the standpoint of the pluralist evaluator – it is a story of tempered predictivism of a sort that applies to virtuous endorsers. But it can be retold in terms amenable to the individualist evaluator. The individualist evaluator, again, simply assesses all relevant evidence for herself – in the case involving Peter, she will possess all evidence possessed by Peter and note that T is, given the relevant evidence, sufficiently well supported to merit some particular degree of endorsement without appeal to the truth of N (recall that Peter is virtuous – and we assume for simplicity that Eva knows nothing relevant unknown by Peter). Thus the 'prediction per se' in this case amounts to the existence of the independent reasons R (which do not include N) that support T to this degree – success predictivism amounts to the fact that R is subsequently supported by the determination of N's truth in this scenario. In the scenario involving Alex, T is simply less well supported by the total available evidence (which we assume, again for simplicity, is simply the total evidence possessed by Alex, possessed also by Eva) – thus no prediction per se is made – and thus any reasons possessed by Alex are not confirmed by the subsequent determination of N's truth. Thus there is no success predictivism. Hereby the individualist evaluator discerns an instance of thin predictivism for the case of virtuous endorsers.

The difference between per se predictivism and success predictivism – in both the tempered and thin forms of virtuous predictivism – can be understood as a difference between the relationships between R and N in the prediction scenario. In the case of per se predictivism, the relationship is combinatorial: the confirming power of R – whatever Eva estimates it to be, given whatever preliminary confidence she has in it – simply combines in Eva's thinking with the confirming power of N to produce an overall higher probability for T in the prediction scenario. In the case of success predictivism, the relationship is revisionary – the success of the prediction induces Eva to revise her estimation of the probability of R. Under what conditions does virtuous predictivism hold? The answer depends on whether one considers the tempered form or the thin form of predictivism. In the tempered form, per se predictivism holds because the act of prediction carries evidence for Eva of the existence of some reason for the theory that Eva does not already possess and would, in her estimation, raise her probability for T if she possessed it. Tempered per se predictivism thus fails when Eva is convinced, for whatever reason, that no such reason exists. This could occur when Eva is convinced that Peter's expertise, however great, is substantially less than hers – such that it is impossible that he could possess a reason not possessed by her. Tempered per se predictivism may fail when the theory endorsed by Peter is one that Eva would, given her own knowledge of the field, have endorsed herself – for the act of endorsement in this case may carry no evidence of a reason that Eva does not possess, as Peter may be constructing T on the basis of reasons possessed by Eva. Thin per se predictivism, as explained above, holds when the evaluator notes Peter's predictive act to be a product of the existence of Peter's reasons for endorsing T, reasons which the evaluator both knows and assesses herself – reasons which would be confirmed by N if it were shown true. Such reasons are noted to be absent in the case of Alex.

Tempered success predictivism holds in cases in which tempered per se predictivism holds and the predictive success of T confirms the truth of Peter's reason(s) R for T for Eva. This can only occur when Eva is convinced (by Peter's predictive act) of the existence of such R and when she is antecedently uncertain whether R is true. Thin success predictivism occurs in cases in which thin per se predictivism holds and Eva notes that the success of the prediction confirms R. Since per se predictivism requires the existence of reasons that support T in the predictor's scenario, and success predictivism works by confirming the truth of such reasons, success predictivism can and will hold only when per se predictivism holds as well (assuming the reasons are not antecedently certain and thus can have their probabilities raised). This is true of both the tempered and thin forms.

In this section we have developed a theory of predictivism that can be used by evaluators who are confronted by endorsers who are assumed to be virtuous, and noted the form this theory takes in its tempered and thin versions. But a more precise explication of this entire approach to predictivism outlined in this section is seriously needed. To this task we now turn.

3.4 VIRTUOUS PREDICTIVISM IN BAYESIAN TERMS

Hypothesis support, we are often told, is a three-way relation (e.g., Good 1967). There is no fact of the matter what probability should be assigned to some hypothesis H on the basis of some evidence E except relative to some body of background belief K. The clear implication of this claim is that, at least in some cases, there is relative to some K a fact of the matter what probability should be assigned to T on the basis of E – and I will adopt this assumption in what follows.[3] This commonplace about three-way support bears importantly on how various issues discussed above should be

---

[3] There are clearly some cases in which this assumption is straightforwardly justified and others in which it is not. If K states that one has purchased a ticket in a fair lottery that includes 100 tickets, then

understood. To begin with, consider the nature of the virtuous poster of probabilities – the agent who posts a probability for a theory that is 'actually justified' on the basis of the evidence he possesses. A virtuous poster, then, is an agent who posts a probability for a theory on the basis of some evidence he does in fact hold that is, given his actual background belief, actually justified.

Let us consider again the point – in each scenario – in which either Peter or Alex posts a prior probability for T (i.e. a probability on the basis of O and whatever other reasons they may possess, prior to considering the impact of N). Alex's prior A1 is non-0 but sub-endorsement, while Peter's prior P1 is an endorsement level probability. We are assuming that both agents are virtuous – they post the probability for T on O that is actually justified for them. But why do they post different probabilities? The obvious explanation is that they are possessed of distinct bodies of background belief. These background beliefs will include what I referred to as 'reasons' in the previous section. I will make a number of assumptions about how Eva views each theorist in the two scenarios described above. I will assume, as noted, that Eva regards each theorist as competent. Given this assumption she assigns to each in the relevant scenario at the outset a probability that his background beliefs are true – the probability is greater than 0 and less than 1. (I follow many others in choosing not to assign probability 0 or 1 to contingent statements.) Let Kp and Ka be the background beliefs of Peter and Alex respectively – in each scenario Eva initially regards the probability of the theorist's background beliefs as having one and the same value S. In neither scenario does she confront both theorists or assign probabilities to both Kp and Ka – only to one or the other. The question is – how do Eva's posterior probabilities for T compare at the end of the prediction and accommodation scenarios?[4]

---

one is justified given K in assigning the probability 0.01 to the proposition that one will win. In some other cases, K may include too little information and thus may not strictly determine a point-valued probability for T. In such cases, everything of substance that transpires below may be preserved if the relevant endorser or evaluator endorses not a point valued probability but an interval valued probability, though the formal analysis becomes messier. For simplicity I retain the assumption that the relevant K justifies a point-valued probability for T in all examples that follow – an assumption adopted in much Bayesian literature.

[4] I am not assuming that the background beliefs that constitute Kp and Ka exhaust the doxastic commitments of the predictor and the accommodator. I do not, in particular, take these background beliefs to be the complete probability functions of the theorists. The relevant K consists of just the portion of the theorist's probability function which is, first of all, relevant to the case at hand, and which is defined on propositions to which probabilities sufficient for belief, in some ordinary sense, are assigned. I am assuming that the remainder of the theorist's relevant probability function can be computed from this subset of the function.

To settle this question, a number of other issues must be settled first. What about Eva's own background beliefs, Ke? Ke entails in each scenario that the probability of the relevant endorser's background beliefs is S – but what else constitutes Ke? We could think of Eva in any of several ways: Eva may also assign high probability to a set of background beliefs K' that constitute a fully worked out view of the relevant domain that is incompatible with the background beliefs of the endorser – perhaps Ke entails the disjunction K' v Kp (in scenario 1) and K' v Ka (in scenario 2) and entails that $p(K') = .95$, with Kp (in 1) and Ka (in 2) having probability S = 0.05. K' could constitute the 'reigning orthodoxy' (e.g., the Newtonian world-view in 1905) and Kp (or Ka) some newly emerging view (e.g., the then emerging Einsteinian worldview). But one could also think of Eva as possessed of no theoretical views of her own to disjoin with Kp or Ka – Eva could play the role of a modestly skeptical non-expert who evaluates a candidate expert's theory – and regards the probability that the candidate has true background beliefs as having some non-negligible value S though she has no theoretical views of her own. In another type of case, Eva could play the role of a trusting non-expert and regard the background beliefs of the endorser as highly probable and complete relative to the state of current scientific knowledge, and thus be nearly prepared to take over that endorser's probability as her own. (We consider various differences between expert evaluators and non-expert evaluators below.)

In what follows I will refer to Eva's prior and posterior probabilities as $p(T/Ke)$ and $p(T/N \& Ke)$ respectively. These are the probabilities that incorporate both her own background views K' and her judgment that the background beliefs of the endorser (either Kp or Ka) has probability S. But at various points it will be important to refer to $p(T/K')$ and $p(T/N \& K')$ – these I will refer to as Eva's individual prior and individual posterior for T. These are the probabilities that incorporate only the background theory K' to which she is reasonably committed, i.e., her own views about the nature of the relevant domain (so if she is a non-expert, K' has little or no content). One example of the importance of Eva's individual probabilities is the following point: it is important that Eva's individual posterior for T not be greater than the accommodator's posterior, and hence certainly less than the predictor's posterior. This is because it is crucial to the thought experiment that, in the accommodation scenario, she view Alex as an endorser – else she would not view him as an accommodator (for an accommodator is a particular type of endorser). But if her individual posterior were greater than his, she could not view him as an endorser. This is because, as noted in (2.2), for an agent to qualify for an evaluator as an endorser, the evaluator's probability for T must not be less than the agent's.

I am assuming that all three bodies of background belief – the predictor's, the accommodator's, and the evaluator's, do not include any knowledge of N until the point at which N is presented to the relevant agent in the thought experiment. This assumption, once again, is not essential: it could be the case that any or all of the three agents know N prior to the thought experiment, but if this is the case, the relevant body of background belief must be purged of N. (For detailed instructions about how to purge a body of background belief of some empirical belief (and thus for a solution to the quantitative problem of old evidence) see Chapter 7.) This is because, for expository purposes, we want the distinction between the agent's prior and posterior probabilities to reflect the evidential impact of the observation that N is true.

In both the prediction and accommodation scenario, Eva takes O as background knowledge, as do Peter and Alex. O is evidence that is accommodated by both the predictor and the accommodator – while the accommodator, but not the predictor, also accommodates N. The critical assumption is that T is more probable on Kp (given that O is background knowledge) than it is on Ka – since Peter's prior (P1) is at or above some contextually fixed limit L for endorsement, while Alex's (A1) is not. We will use 'v1' to denote Eva's probability function in scenario 1, and 'v2' to denote her function in scenario 2. $v1(T/Kp)$ is thus equal to Peter's prior probability P1 and $v2(T/Ka)$ is equal to Alex's prior probability A1. Thus we assume:

$$v1(T/Kp) \geq L > v2(T/Ka) > 0. \qquad (1)$$

Our objective is to determine whether $v1(T/N)$ is greater than $v2(T/N)$, i.e., whether there is a predictivist effect for Eva. I will show that there is, under several very reasonable assumptions – and then show that our computations neatly reflect the two roots of predictivism described informally in the previous section.

In what follows I will develop two thought experiments – both of which reveal a predictivist effect. In both experiments, Peter is the predictor and Alex the accommodator – and the various relevant facts about Peter and Alex are otherwise kept as similar as possible. However, there is more than one way in which facts can be kept maximally similar in two scenarios. In the first thought experiment, Peter and Alex are assumed to endorse the same value for the likelihood ratio $P(N/T)/P(N/\sim T)$ – the effect of this is that their posterior probabilities for T are different in a way that reflects the difference in their prior probabilities (cf. (1) above). In the second thought experiment, Peter and Alex are assumed to have the same posterior probability for T – consequently their values for the likelihood ratio differ. But the result in both experiments is the same: there is a predictivist effect for Eva.

3.4.1 The first thought experiment

I propose to make the following assumptions to the effect that there is no non-arbitrary difference between probabilities as assigned in the two scenarios – I will refer to 2–7 as the ceteris paribus assumptions:

$$v_1(T/N \& \sim Kp) = v_2(T/N \& \sim Ka) = C \tag{2}$$

$$v_1(N/\sim T \& Kp) = v_2(N/\sim T \& Ka) = C\# < 1 \tag{3}$$

$$v_1(N/\sim Kp) = v_2(N/\sim Ka) = C^* \tag{4}$$

$$v_1(N/T \& Kp)/v_1(N/\sim T \& Kp) = v_2(N/T \& Ka)/ \\ v_2(N/\sim T \& Ka) \tag{5}$$

$$v_1(T/N \& Kp) > C \tag{6}$$

$$0 < v_1(Kp) = v_2(Ka) = S < 1 \tag{7}$$

(2) reflects my assumption that Eva's reasons for thinking that T is true (based on K') – assuming N – other than those provided by the endorser's background beliefs – are the same in the two scenarios. In other words, Eva's individual posterior is the same in the two scenarios. This seems fair, as any difference between the two scenarios in this respect would simply be arbitrary. (Matters would be different, of course, if Eva encountered both the predictor and the accommodator side by side in a single scenario – then the probability of T on N assuming the negation of one set of background beliefs would depend on the probability as determined by the other set and (2) would not hold – but this is not a scenario we are considering.) (3) reflects the assumption that Kp and Ka do not differ with respect to the reasons they provide for N other than its being a consequence of T, as any such difference would be arbitrary. If C# were not less than 1 then Kp and Ka would entail N (since T entails N), contrary to our assumption that Kp and Ka do not contain the information that N is true. (4) reflects the

assumption that whatever reasons Eva may have for thinking N true (based on K'), apart from those provided by Kp or Ka, do not vary between the two scenarios (i.e. Eva's individual priors for N are the same in both scenarios). (5) asserts that the degree of confirmation provided by N to T is the same for the two endorsers – assuming degree of confirmation is a function of the likelihood ratio (see Fitelson 1999 S363 for a list of many philosophers who endorse this confirmation measure). (Note that (3), together with the fact that T entails N, entails (5); (3) and (5) will be dropped in the second thought experiment.) (6) reflects the requirement that Eva's individual posterior cannot be greater than Alex's posterior: if (6) were not true, then Eva's individual posterior would be no less than Peter's posterior, and thus it would be greater than Alex's in the accommodation scenario, and thus she could not view Alex as an endorser. As noted above, the conditional probabilities in (2) are equivalent to Eva's individual posterior for T – they are the probability of T on N assuming just K'. So Eva's individual posterior for T should not be too high. At the same time, neither should this probability be thought of as taking o as a possible value – for neither is Eva a dogmatic repudiator of T. Her individual posterior for T on N must fall within some neutral range, establishing her as neither an endorser nor an absolute repudiator of T (it is possible that she might qualify, with Alex, as an accommodator, if her individual posterior were the same as Alex's posterior). This is not to say, of course, that Eva's posterior probability itself (taking into account the information provided by Peter or Alex) must be at a sub-endorsement level in either scenario. (7) embodies the assumption that Eva begins both scenarios regarding the background beliefs of the endorser as equiprobable.

With assumptions 1–7 in hand, the demonstration that $v_1(T/N) > v_2(T/N)$ is straightforward. Eva's posterior probability – her probability for T on N – in the prediction scenario can be expanded by the theorem of total probability as follows:

$$v_1(T/N) = v_1(T/N \& Kp)v_1(Kp/N) + v_1(T/N \& \sim Kp)$$
$$[1 - v_1(Kp/N)] \tag{8}$$

Her posterior probability for T in the second scenario – the accommodation scenario – is likewise as follows:

$$v_2(T/N) = v_2(T/N \& Ka)v_2(Ka/N) + v_2(T/N \& \sim Ka)$$
$$[1 - v_2(Ka/N)]. \tag{9}$$

Re-writing (8) and (9), given (2), and re-arranging, we have

$$v_1(T/N) = v_1(Kp/N)(v_1(T/N \& Kp) - C) + C \qquad (10)$$

$$v_2(T/N) = v_2(Ka/N)(v_2(T/N \& Ka) - C) + C \qquad (11)$$

Recall at this point (6) which assures us that $(v_1(T/N \& Kp) - C)$ is positive. Our desired result – that $v_1(T/N) > v_2(T/N)$, follows immediately if (A) and (B) below can be shown:

$$v_1(T/N \& Kp) > v_2(T/N \& Ka) \qquad (A)$$

$$v_1(Kp/N) > v_2(Ka/N). \qquad (B)$$

And these can both be shown quite easily. (A) asserts that Peter's posterior probability is greater than Alex's, and this follows from the conjunction of (1) and (5) – Peter's posterior probability must be greater than Alex's since Peter's prior is greater and their respective likelihood ratios are identical. (B) follows by the following computations:

$$v_1(Kp/N) = \frac{v_1(Kp)v_1(N/Kp)}{v_1(N/Kp)v_1(Kp) + v_1(N/\sim Kp)[1 - v_1(Kp)]} \qquad (12)$$

$$v_2(Ka/N) = \frac{v_2(Ka)v_2(N/Ka)}{v_2(N/Ka)v_2(Ka) + v_2(N/\sim Ka)[1 - v_2(Ka)]} \qquad (13)$$

Bearing in mind (4) and (7), we have:

$$v_1(Kp/N) = \frac{Sv_1(N/Kp)}{v_1(N/Kp)S + C^*(1 - S)} \qquad (14)$$

$$v_2(Ka/N) = \frac{Sv_2(N/Ka)}{v_2(N/Ka)S + C^*(1 - S)} \qquad (15)$$

One can re-write (14) as follows:

$$v_1(Kp/N) = \frac{1}{1 + C^*(1 - S)/Sv_1(N/Kp)} \qquad (16)$$

(16) follows by dividing the numerator and denominator of (14) by $Sv_1(N/Kp)$ (recall that S is greater than 0, as is $v_1(N/Kp)$, since $v_1(T/Kp) > 0$ and N is a logical consequence of T). A similar move allows us to re-write (15) as follows:

$$v_2(Ka/N) = \frac{1}{1 + C^*(1 - S)/Sv_2(N/Ka)} \qquad (17)$$

But clearly $v_1(Kp/N)$ will be greater than $v_2(Ka/N)$ (and thus (B) will be true) if $v_1(N/Kp)$ is greater than $v_2(N/Ka)$. But this follows quickly:

$$v_1(N/Kp) = \frac{v_1(N/T \,\&\, Kp)v_1(T/Kp) + v_1(N/{\sim}T \,\&\, Kp)}{[1 - v_1(T/Kp)]} \qquad (18)$$

$$v_2(N/Ka) = \frac{v_2(N/T \,\&\, Ka)v_2(T/Ka) + v_2(N/{\sim}T \,\&\, Ka)}{[1 - v_2(T/Ka)]} \qquad (19)$$

Recall that N is a logical consequence of T, so $v_1(N/T \,\&\, Kp) = v_2(N/T \,\&\, Ka) = 1$.

Remembering (3), we can re-write (18) and (19), and re-arrange, thus:

$$v_1(N/Kp) = v_1(T/Kp)[1 - C\#] + C\# \qquad (20)$$

$$v_2(N/Ka) = v_2(T/Ka)[1 - C\#] + C\# \qquad (21)$$

Since $0 \leq C\# < 1$ (by (3)), it follows that $v_1(N/Kp) > v_2(N/Ka)$ iff $v_1(T/Kp) > v_2(T/Ka)$ – but we assumed the latter in (1). Thus $v_1(N/Kp) > v_2(N/Ka)$, from which it follows that $v_1(Kp/N) > v_2(Ka/N)$ – thus we have shown (B) above. From the conjunction of (A) and (B) it clearly follows that $v_1(T/N) > v_2(T/N)$ – thus Eva regards T as more probable in scenario 1 than in scenario 2 – and thus predictivism holds true for Eva on assumption (1), together with the ceteris paribus assumptions (2)–(7).

The formal analysis provided in this section dovetails with the informal analysis provided in the preceding section. There, I argued that the intuition that Eva's posterior probability is greater in the prediction scenario than in the accommodation scenario had two roots. On the one hand, the prediction per se – the endorsement of T without appeal to knowledge of N – on Peter's part testified to Peter's possession of reasons for T that were not possessed by Alex. The stronger that endorsement, the greater is the difference between Peter's and Alex's priors, and thus the greater will be the predictivist effect. On the other hand, the predictive success of Peter's predictions supported the claim that Peter's reasons for endorsing T were true. The greater that support, the greater is the predictivist effect. The formal counterparts to per se and success predictivism can be found, essentially, in the claims (A) and (B) which were shown to hold true in the above analysis and which grounded the predictivist effect. (A) claims that Peter's posterior probability is greater than Alex's – this was, given the equivalence of Peter's and Alex's likelihood ratios, an immediate result of Peter's prior being greater than Alex's. But it was Peter's enhanced prior that constituted Peter's status as an endorser of T and as a predictor of N. The prediction per se was identical to his posting of the endorsement level probability that reflected his reasons for believing T. That Eva's posterior probability was in part a function of Peter's prior reveals the epistemic significance, for Eva, of Peter's prediction per se – and the sensitivity of Eva's posterior probability to the strength of that predictive act. As to (B), it reveals that the predictivist effect is attributable in part to the success of Peter's prediction, the demonstrated truth of N, to confirm Kp more strongly than it confirms Ka. But the effect of the enhanced probability of the predictor's background beliefs is that the predictor's reasons for endorsing T and thus predicting N are shown to be more likely to be true than in the case of the accommodator. This is success predictivism. Thus the two roots of predictivism are nicely illustrated in the sensitivity of the predictivist effect to the inequalities (A) and (B).[5]

The analysis of the distinction between the thin and tempered versions of virtuous predictivism laid out in the previous section applies to the content of this section as well. The probabilistic computations are the same

---

[5] A possible objection to my claim that (A) and (B) reflect two distinct roots of virtuous predictivism is that (A) and (B) manifest not two distinct roots but simply one. This is because in the thought experiment presented above, the truth of (B) seems to derive in a straightforward way from the truth of (A). After all, the only reason N confirms Kp more than Ka is that v_1(N/Kp) > v_2(N/Ka) – but this inequality results just from the fact that v_1(T/Kp) > v_2(T/Ka) (given that T entails N, and given the ceteris paribus assumptions). Thus the so-called 'root' of the success effect appears more or less identical to the root of the per se effect. One response to this objection is to note that from the standpoint of an evaluator like Eva, the distinction between the two roots is particularly palpable given that she can experience their effects at two distinct points in time – that is, if N is shown true after Peter's prediction. The per se effect takes hold at the moment of Peter's prediction, the success effect only later when N is shown true. If there were at the heart of things just one root, how could its effects occur at separate points in time? Another response is to concede that the appearance of a single root is an artifact of the ceteris paribus assumptions that drive this particular example – but other examples can be presented in which the distinction between the per se and success roots is quite salient. Suppose that, as before, Peter's prior for T is endorsement level but Alex's is not – but in this pair of scenarios, Eva queries Peter and Alex as to how probable N is on their respective background beliefs. Surprisingly, v_1(N/Kp) = v_2(N/Ka) – this equality reflects, let us suppose, that the likelihood ratios (p(N/T)/p(N/~T) are not the same for Peter and Alex (contrary to assumption (5) above). The fact that v_1(Kp) = v_2(Ka) thus entails that v_1(Kp/N) = v_2(Ka/N). In this event, Alex's likelihood ratio is lower than Peter's, so N is more probable on ~T for Alex than it is for Peter – thus the probability of N on Ka manages to be equal to the probability of N on Kp. This means that the disposition of N to confirm Kp is exactly the same as N's disposition to confirm Ka – and so there is no difference between the two scenarios at the level of background belief confirmation. There will remain, however, a predictivist effect for Eva that derives from the fact that v_1(T/Kp) > v_2(T/Ka) – the per se effect. Thus the per se effect can operate without the simultaneous operation of the success effect – so the claim that the two effects are really just one is refuted. (But as noted above, the success effect depends upon the presence of the per se effect – if there is no difference between Peter's and Alex's prior probabilities, for one thing, it is impossible for one of them to qualify as a predictor while the other does not.)

whether Eva is a pluralist, and assigns probabilities based on her willingness to trust Peter and Alex and their respective background beliefs, and various probabilities, or an individualist, and assigns probabilities based entirely on her own assessment of the relevant evidence. The Bayesian analysis thus illuminates the two roots of virtuous predictivism in its thin and tempered forms.

3.4.2 The second thought experiment

Despite its many logical nuances, it is not surprising that Eva experiences a predictivist effect in the first thought experiment – after all, she is a pluralist who is confronted with endorsers she knows to be virtuous, and Peter posts a higher prior for T than does Alex (this is what accounts for the fact that Peter is a predictor and Alex an accommodator). Given the assumption that Peter and Alex attach the same value to the likelihood ratio, it follows that Peter's posterior probability is higher than Alex's – so it is little wonder that Eva's posterior probability for T is higher in the Peter scenario. I have tried to show, however, that a full account of the predictivist effect on these assumptions is more complex and interesting than this simple account might suggest. But now let us consider a different thought experiment – one in which Peter posts the same prior he did in the first, and Alex does as well, but in which incoming knowledge of N serves to drive both their posteriors to a common posterior value. This means, of course, that N will drive Alex's probability for T farther up than N drives Peter's probability – which entails that Alex's likelihood ratio ($P(N/T)$/ $P(N/\sim T)$) is higher than Peter's likelihood ratio. Now Eva is confronted with scenarios in which Peter and Alex endorse a common posterior for T, though Peter endorsed T without appeal to N and thus predicted N, while Alex did not. We retain all assumptions (1)–(7) except for (3) and (5). Is there a predictivist effect for Eva? I will argue that there is.

However, there is another purpose in considering this second thought experiment, and it is to bring to light an interesting feature of the first. The computations of the first thought experiment begin with (1) ($v_1(T/Kp) \geq L > v_2(T/Ka) > 0$). The role of this assumption in the computations of the first experiment, however, is just that it insures that Peter's prior for T is greater than Alex's, and that Eva's own prior probabilities are not greater than Alex's posterior probability (so that Eva can view Alex as an endorser). From some meager assumptions, a predictivist effect is shown to follow. However, it should be remembered that there are actually two conditions on L, as explained in (2.2): (1) a T-endorser's probability for T cannot be less than an evaluator's probability for T, and (2) an endorser who posts a probability for T that is at or above L is one whose credibility is enhanced, for a pluralist evaluator, by subsequent confirmation of some empirical consequence N of T. Thus an endorser who posts such a probability qualifies as a predictor of N. An endorser who posts a probability for T that is below L is one whose credibility is not enhanced by subsequent confirmation of N, and thus fails to predict N. The first thought experiment makes use of the first condition on L but not the second. In fact, the second condition is neglected in the first experiment to make the point that it is not needed to demonstrate a predictivist effect. However, we will need to make use of the second condition in this second thought experiment.

Let v^*_1 refer to Eva's probability function in the predictor's scenario in the second thought experiment (it is identical to v_1, as the predictor scenarios are identical in the first and second experiments). Let v^*_2 refer to Eva's probability function in the accommodator's scenario in the second experiment – this will incorporate the new assumption that Alex's posterior probability for T is the same as Peter's. For this experiment, we will need to make use of the fact that the basic concept of L thus makes the following claims true (for a pluralist evaluator like Eva) essentially by definition of L:

$$v^*_1(Kp/N) > v^*_1(Kp) \tag{22}$$

$$v^*_2(Ka/N) \leq v^*_2(Ka) \tag{23}$$

Assuming, with (1), that $v^*_1(Kp) = v^*_2(Ka)$, it follows that

$$v^*_1(Kp/N) > v^*_2(Ka/N). \tag{24}$$

But with (24) in hand the demonstration of a predictivist effect for Eva (viz., that $v^*_1(T/N) > v^*_2(T/N)$) in the second experiment is straightforward. We begin as before by expanding Eva's two posterior probabilities by the theorem of total probability:

$$v^*_1(T/N) = v^*_1(T/NKp)v^*_1(Kp/N) + v^*_1(T/N \sim Kp) \atop (1 - v^*_1(Kp/N)) \tag{25}$$

$$v^*_2(T/N) = v^*_2(T/NKa)v^*_2(Ka/N) + v^*_2(T/N \sim Ka) \atop (1 - v^*_2(Ka/N)) \tag{26}$$

Re-arranging, we obtain

$$v^{*}1(T/N) = v^{*}1(Kp/N)[v^{*}1(T/NKp) - v^{*}1(T/N \sim Kp)] \\ + v^{*}1(T/N \sim Kp) \tag{27}$$

$$v^{*}2(T/N) = v^{*}2(Ka/N)[v^{*}2(T/NKa) - v^{*}2(T/N \sim Ka)] \\ + v^{*}2(T/N \sim Ka) \tag{28}$$

Our new assumption, however, is that $v^{*}1(T/NKp) = v^{*}2(T/NKa)$ (i.e. Alex and Peter have a common posterior probability); also, $v^{*}1(T/N \sim Kp) = v^{*}2(T/N \sim Ka)$ (by (2)). But now it suffices to show that $v^{*}1(Kp/N) > v^{*}2(Ka/N)$ – but this has been shown already as (24). It follows that $v^{*}1(T/N) > v^{*}2(T/N)$ – there is a predictivist effect in the second thought experiment.

Alex's endorsement of T, when it comes upon the revelation of N, is stronger in the second thought experiment than in the first. But intuitively, the fact that Alex's posterior soars to the level of Peter's offers little assurance to Eva in the accommodator's scenario – for it comes too late. The fact that Alex's prior was sub-endorsement means that the subsequent demonstration of N offers no indication of his credibility. This is the conclusion of the second thought experiment, though it is important to recall that the result in this case was reached by appealing critically to (22) and (23) which were based simply on the basic concept of L.

The predictivist effect in the second thought experiment is somewhat more surprising than in the first, and illustrates that the phenomenon of virtuous predictivism is a robust one.[6]

3.5 FOUR SPECIES OF PREDICTIVISM

Our analysis at this point has postulated a basic distinction between thin and tempered predictivism, and also a basic distinction between virtuous and unvirtuous predictivism. These distinctions intersect to produce four species of predictivism. We assume, as usual, that T entails true N:

---

[6] CUP referee B worries that the virtuous predictivist effect is 'under threat' if it is the case that the confirmation prediction provides is simply by means of confirming T, which in turn confirms the background. In response I would say that in the typical case, N (the novel consequence of T) does confirm Kp simply because N confirms T. This is because for the predictor, p(T/Kp) is high, and thus p(N/Kp) is high because N is a logical consequence of T (it is important that p(N/Kp) NOT be high because there is some other theory T*, which Kp supports, that entails N (cf. 2.2 ft.3). The high value of p(N/Kp) is part of what makes N confirm Kp so highly, and it is attributable to the fact that p(T/Kp) is high (compared to the accommodator's p(T/Ka)). So it seems to me correct to say that N confirms Kp because it confirms T, though this locution is somewhat obscure.

1. Unvirtuous thin predictivism: an individualist evaluator notes that an accommodator has posted an endorsement probability for T on a basis that includes N in a way that incorporates a logical misstep (often the result of social pressure to endorse an N-entailing theory). Alternately, the evaluator notes that a predictor has endorsed T without appeal to N (and thus avoided any N-induced logical misstep). The individualist evaluator, insofar as she is also an unvirtue-ascribing evaluator, thus discerns a correlation between accommodation and the lack of evidence for T, viz., the lack of evidence that would legitimate the accommodator's endorsement level probability for T. Equivalently, there is a correlation between prediction and the existence of evidence for T, viz., the existence of evidence that legitimates the predictor's endorsement of T.

2. Unvirtuous tempered predictivism: a pluralist evaluator takes an accommodator's endorsement of T on a body of evidence that includes N as genuine evidence of an irrationally inflated probability for T, given the accommodator's presumed desire to endorse an N-entailing theory and given the possibility that the endorser is unvirtuous. Alternately, the pluralist takes the successful prediction of N by a T-endorser as evidence of the non-existence of such an irrationally inflated probability – given that ignorance of N was one less motive for adhocery.

3. Virtuous thin predictivism: an individualist evaluator notes the fact that some set of background beliefs K led an endorser to endorse T without appeal to N. K offers reasons for believing T (per se predictivism), and the truth of N serves to confirm K (success predictivism). Alternately, the individualist notes that these two forms of support are not available for some K which led some accommodator to endorse T on a basis that included O and N.

4. Virtuous tempered predictivism: a pluralist evaluator takes a prediction by a virtuous endorser as evidence of the existence of some K that supports T (which motivated the endorser to endorse T) and takes the success of that prediction to confirm K. Alternately, the pluralist does not find an accommodation by a virtuous evaluator to provide similar support.

This four-way distinction reveals that the question whether predictivism holds true for some scientific community is too simple – the question ought to be which, if any, of the four species of predictivism identified here characterizes theory assessment in some scientific community. My position, suggested to some extent already but expounded more fully below, is that a typical scientific community of evaluators will be a diverse lot, and that there is nothing unusual about a community containing evaluators which instantiate some or all of the various types. I will argue now that this is the case in a particularly famous episode of theory evaluation.

3.6 MENDELEEV AND THE PERIODIC LAW

The Periodic Table of the Elements is presented in every chemistry textbook. The scientist most famously credited with discovery of this table is the Russian chemist Dmitri Mendeleev (1834–1907), who proposed an early version of this table in papers published in 1869 and 1871 and in a textbook consisting of multiple volumes published over several years, beginning in 1869. The theoretical content of his Periodic Table is constituted by the periodic law (PL), which asserts a nomological relationship – expressed by a periodic function – between the atomic weight of each element and its various physical and chemical properties.[7] The example of PL is relevant in our present context because the story of its eventual acceptance by the scientific community seems, to many commentators, to be a spectacular example of predictivism in action. Mendeleev showed the known elements (numbering over 60) could be accommodated by his version of PL, which ordered the elements according to their atomic weight and showed how many chemical and physical properties recurred periodically. But he also made several bold predictions – including that three gaps in his table would be eventually filled by the discovery of new elements, and that various values for atomic weights would be changed to accord with his estimations of their values. The three elements – which were eventually named gallium, scandium, and germanium – were discovered in 1875, 1879, and 1886 respectively. By the end of this period two of the atomic weight predictions had been fulfilled as well. Mendeleev was awarded the prestigious Davy Medal by the Royal Society in 1882 for his work on PL – after gallium and scandium were discovered and one weight correction (for uranium) was confirmed.

Many authors who have addressed the question have agreed that the success of Mendeleev's predictions confirmed the PL more strongly than a similar number of accommodations, viz., a similar number of elements whose weights and properties were written into the PL by Mendeleev. As Maher writes, "The only plausible explanation [for Mendeleev's winning of the Davy Medal] is that scientists were impressed by the fact that these latter pieces of evidence were verified predictions rather than accommodated

---

[7] Of course, contemporary versions of the Periodic Table portray physical and chemical properties of elements as a function not of their atomic weight but of their atomic number.

evidence." (1988: 275) Likewise Lipton summarizes the scientific community's assessment of Mendeleev's evidence by claiming that "Sixty accommodations paled next to two predictions" (1991: 134). Very detailed and compelling historical evidence for this conclusion is provided by Brush's (1996). Brush's evidence takes the form of a scholarly study of chemistry textbooks in Great Britain and the United States in the late nineteenth century. He concluded that "Mendeleev's periodic law attracted little attention (at least in America and Britain) until chemists started to discover some of the elements needed to fill gaps in his table and found that their properties were remarkably similar to those he had predicted. The frequency with which the periodic law was mentioned in journals increased sharply after the discovery of gallium; most of that increase was clearly associated with Mendeleev's prediction of the properties of the new elements." (1996: 617)

Scerri and Worrall, in their 2001 article entitled "Prediction and the Periodic Table," provide an in-depth historical analysis of the facts surrounding the scientific community's evaluation of PL.[8] Scerri and Worrall argue that there are reasons to be skeptical about some of the reasons cited in defense of the claim that PL's successful predictions played a significantly more important role than its accommodated evidence. In response to Maher, Scerri and Worrall point out that the citation scripted by the Royal Society that awarded the Davy Medal to Mendeleev did not even mention the successful predictions (2001: 416–417) (this objection is reiterated in Scerri (2007: 145–146)). In response to Brush, Scerri and Worrall point out that among the many textbook citations of PL cited after Mendeleev's predictions began to be confirmed, many do not mention the successful predictions (2001: 433–434). They take this absence of explicit comments about the importance of successful prediction to count as serious evidence against Brush's (1996) predictivist conclusions. In my view, however, Scerri and Worrall are vulnerable to the charge that they are committing a classic fallacy of social science: they would base their view of what members of a community are doing on what such members say (or fail to say) they are doing, rather than what they are actually doing. Moreover, there is no reason to think that the Davy Medal citation or the textbooks that discuss PL would mention the successful predictions even if these successful predictions were the most important evidence in favor of PL. Authors of such references might well have chosen to focus simply on the content of PL and its importance for the field of chemistry rather than

---

[8] This paper also further articulates and defends Worrall's theory of predictivism (cf. 1.4.4) which is applied to the case of PL; I respond to this application below.

on the reasons for its acceptance. Brush's evidence thus seems to speak incontrovertibly in favor of the claim that predictivism strongly held in this episode. Certainly Scerri and Worrall offer no other explanation for the dramatic increase in references to PL in nineteenth-century chemical text-books with the success of the relevant predictions.

However, Scerri and Worrall proceed to develop another interesting reason for withholding a strongly predictivist interpretation of the story of PL's evaluation by the scientific community. The relevant story involves the process by which the noble gases and rare earths came to be accom-modated into PL. Scerri and Worrall depict the story of the noble gas argon as typical: discovered in 1895 by Rayleigh and Ramsay, argon's estimated atomic weight did not accord it any obvious position in the Periodic Table. Argon thus emerged as an anomaly for PL, and at least one prestigious scientist (the president of the Royal Society) pointed out that it might serve to refute PL. But ultimately the anomaly was resolved when it was pro-posed that argon could be accommodated by incorporating a new group in the Periodic Table. Scerri and Worrall assert that "the accommodation of argon within Mendeleev's scheme was a major feather in its cap – no less major than any other empirical success, whether predictive in the tem-porally novel sense or not." (2001: 445) So at least in some cases, they conclude, evidential accommodation was at least as important as successful prediction for the ultimate acceptance of PL. However, Scerri and Worrall go on to explain what this evidential accommodation actually involved: "It is not simply a question of inventing a new section of the table to fit the noble gases into; it must then also be checked that the periodicities previously noted in terms of valencies, 'analogous properties' and the like among the already accommodated elements are preserved." (2001: 445) In other words, once the new spaces for the noble gases were invented, scores of new predictions issued from these inventions which had to be checked against the facts. Scerri and Worrall conclude that "creating a new group for the noble gases leads to a new series of predictions (in the atemporal sense) about already known analogies between elements." (2001: 446) But if this is the case, why not understand the importance of the 'accommo-dation' of the noble gases and rare earths as primarily or perhaps entirely grounded on the success of these predictions? The story of argon provides little ammunition against predictivism here after all. I conclude on the basis of all these considerations that predictivism clearly held in the scientific community's evaluation of Mendeleev's theory.

I will argue in what follows that this predictivist effect can be explained in terms of the account of predictivism I have outlined in this chapter.

Before this can be shown, however, a number of preliminary points must be made. For an episode of theory evaluation to be assessed for predictivism, two roles must be instantiated in that episode: there must be one or more evaluators, whose role is to determine what probability to assign to a theory T, and there must be an endorser, who either plays the role of a predictor or accommodator of some evidence E, depending on whether she appeals to E in the act of endorsing T. In the episode at stake there is a large community of evaluators, who constitute whatever faction of the scientific community adopted an opinion about the probability of PL based on the evidence presented by Mendeleev (and others). Mendeleev of course plays the role of the endorser. Now already there is an important difference between this episode and the idealized illustration of predictivism provided in terms of Eva, Peter and Alex – for the latter involved two scenarios, in one of which Eva confronts Peter the predictor, in the other she confronts Alex the accommodator. The point of the two scenario presentation was to illustrate the dynamics of predictivism in terms of a controlled thought experiment: keeping all relevant factors between the two scenarios as similar as possible, and varying only the factor of prediction and either the posterior probabilities (the first thought experiment) or the likelihood ratios (the second thought experiment), a difference in the evaluator's posteriors can be thus ascribed to the prediction factor. But in Mendeleev's case, we have only the predictor's scenario – there is no scenario corresponding to the accommodator's scenario in which the evidence predicted by the predictor is accommodated. The story of Mendeleev is an example of what I referred to in (1.2.2) as a 'glorious successful prediction.' It is thus somewhat unclear how a predictivist effect is to be discerned. One way out of this dilemma would be to suggest that there is a scenario corresponding to the accommodator scenario in the case of Mendeleev, and that it is a counterfactual scenario: it amounts to the scenario that would have obtained if Mendeleev (or his counterpart) had accommodated the evidence he predicted in the actual world. Predictivism asserts then that the endorser's actual posterior is higher than her counterfactual posterior. While this approach is a bit tempting, it suffers from the well-known complexities and confusions involved in assessing truth values of counterfactuals. I will adopt another way out, and that is to discern predictivism entirely in terms of what held in the actual world. In terms of this episode in the history of chemistry, this makes predictivism claim that PL was more strongly confirmed by the success of Mendeleev's predictions than it was confirmed by the truth of a similar number of his actual accommodations – bearing in mind that the type of evidence involved in the two cases was qualitatively quite similar. It is this thesis that the various commentators quoted above seem to have in mind when they discern the truth of predictivism in this case. The connection between the two-scenario thought experiments and this historical application is straightforward: assuming the results of the thought experiments are sound, we expect accommodated evidence to confirm less than predicted evidence in a single scenario like the one involving PL.

The claim that the emergence and evaluation of Mendeleev's PL reveals the truth of predictivism thus is based on an historical picture in which there are two major events: in the first event, Mendeleev presented early versions of his Periodic Table based on a body of accommodated evidence. These early versions, as noted, were presented in papers published in 1869, another in 1871, and in a textbook published over a number of years beginning in 1869. In the second event, Mendeleev's bold predictions about the existence of heretofore undiscovered elements and revisions of accepted values of atomic weights are confirmed – this event began in 1875 with the discovery of gallium and continued over the next several years with more confirmed predictions. I will follow Scerri and Worrall (2001: 413) and refer to the period between the first event and the beginning of the second event as 'Stage 1,' and the period following the second event as 'Stage 2.' Thus Stage 1 is roughly from 1869 to 1875, and Stage 2 from 1875 until whatever point in time the PL is accepted without controversy (for present purposes it is not particularly important what that time was). It appears to have been clearly true that the probability of PL – as determined by the relevant evaluators – increased dramatically upon the arrival of Stage 2, suggesting the existence of a predictivist effect.

We turn now to the project of explaining the predictivist effect in terms of the account of predictivism I have provided here. I argued in the previous section that predictivism can take four different forms. The mere existence of a significant increase in assessed probability after the stage transition does not itself indicate which of these forms prevailed. For example, the increase could be accounted for in thin terms – if the relevant evaluators were individualists and thus regarded acts of prediction as epistemically insignificant in themselves but correlated with some significant factor – or in tempered terms – if the existence and ultimate success of Mendeleev's predictive acts were epistemically relevant for pluralist evaluators. Likewise the effect might be explained in either virtuous or unvirtuous terms. To determine which (if any) of the four species of predictivism identified above may have held among the various evaluators, certain facts having to do with Mendeleev himself must be ascertained. One of these concerns Mendeleev's act of theory endorsement. This act, according to my account, is constituted by the posting of a probability for a theory based on a given body of evidence – a probability that exceeds some minimum L requisite for endorsement and whose value figures critically in the evaluator's assessment. In Mendeleev's case this act was constituted by the publications of the early version of his theory – based on the accommodation of widely accepted empirical claims about element properties and atomic weights – and his obvious confidence in the truth of his theory, a confidence that was revealed in the certainty with which he made his predictions (more below). Another important issue that must be carefully examined concerns the background beliefs Mendeleev brought to the table. For the claim that either virtuous or unvirtuous predictivism held is a claim that bears importantly on the background beliefs of the endorser: according to unvirtuous predictivism, the endorser may be prone to post a probability too high based on accommodated evidence, given his own background beliefs, either because he intentionally fails to respect the content of these beliefs or because of logical error (more precisely, as explained above, his posted probability does not cohere with his relevant beliefs). According to virtuous predictivism, predictivism can be true for two reasons – one of which is that prediction indicates the existence of theory-supporting background beliefs more strongly than accommodation. The other is that successful prediction confirms the background beliefs of the endorser more strongly than accommodation. Before proceeding any further then, the question what Mendeleev's background beliefs were deserves careful consideration.

3.7 MENDELEEV'S BACKGROUND BELIEFS

It is often noted that Mendeleev's reputation as the discoverer of PL is a half-truth at best, for a number of other individuals proposed other versions of PL around the time of Stage 1 (Brush identifies five other scientists with claims to credit for the discovery (1996: 596)). Despite the historical importance of these co-discoverers, Mendeleev was unique among them in one important respect: he alone *boldly* predicted that various gaps in the Periodic Table as he had formulated it would eventually be filled. This boldness was of course a result of the enormous confidence Mendeleev had in this theory. It was in part this great confidence that accounts for the fact that he is usually considered the primary discoverer of PL. My position is that this confidence is attributable to various background beliefs held by Mendeleev that were not held by others, or at least not held as strongly. We now turn to the question of what these background beliefs were, why Mendeleev held them, how they supported PL, and the further question of the extent to which these beliefs were shared by others, including (most importantly for our purposes) the relevant evaluators.

Mendeleev is, first and foremost, a predictor: he endorsed a theory, PL, on an epistemic basis that did not include a certain claim N (a conjunction of his various predictions about the existence of certain elements with their various properties, and about the revisions of various atomic weights) – and thus predicted that N would prove true. But, like most predictors (such as Peter), he also plays the role of accommodator insofar as there is a body of evidence he does appeal to in the act of endorsing PL. This body of accommodated evidence consists of the body of 'established fact' regarding the atomic weights of the known elements and their physical and chemical properties. But this body of evidence, while not universally shared by all relevant evaluators (as we will see), was shared by many of them, though it did not produce in these evaluators the same confidence in Mendeleev's theory that it did in Mendeleev himself. What accounted for this difference was, as I claim above, his confidence in a conjunction of various background beliefs. I identify in what follows three background beliefs held by Mendeleev that enabled his confidence in PL, given the accommodated evidence, and discuss each in turn. I refer to the first of these background beliefs as Km1:

3.7.1 *Km1:* The methods of atomic weight calculation endorsed by Cannizarro (at the 1860 Karlsruhe Conference) are sound; many pre-1860 atomic weight estimations made by other means are inaccurate

Mendeleev's journey toward the development of his Periodic Table began at the 1860 Karlsruhe Conference – it was the first international conference on chemistry. The conference had been called for the purpose of addressing several fundamental questions, including the values of atomic weights. In the years prior to the meeting, there was no agreement as to what the value of atomic weights should be (Laing 1995: 151; Ihde 1961: 83). The most important event of the Conference in this regard was the role played by Stanislao Cannizzaro, who urged the chemical community to adopt the procedure used by Gerhardt based on Avogadro's hypothesis. Cannizzaro's arguments had a profound effect on Mendeleev. Years later Mendeleev wrote:

I consider my participation in the congress of chemists in Karlsruhe, 1860, and the ideas presented there by the Italian chemist Cannizzaro, to be a decisive stage in the development of my ideas on periodic law. I soon noticed that the changes in atomic weight that he proposed brought a new harmony to Dumas' groupings, and from this moment on I had the intuition of a possible periodic for the elements ordered by increasing atomic weights. However, I was stopped by the inaccuracy of the atomic weights accepted at that time: the only conviction that remained clearly established was that it was necessary to continue working in this direction. (Quotation given in P. Kolodkine (1964) *DI Mendeleev et la loi periodique*, Paris, Seghers, quoted in Bensaude-Vincent 1986: 8, translation by Martha Nichols-Pecceu).

During the 1860s, many new estimations of atomic weight values that Mendeleev regarded as soundly based on Cannizzaro's proposals were made, and he would later write that at the end of the 1860s "the numerical value of atomic weights became definitely known" (Mendeleev 1889: 324). These were the estimations that Mendeleev would use as accommodated evidence in his first published version of the Periodic Table in 1869 – the beginning of Stage 1.

Obviously, Mendeleev's strong commitment to Km1 played a critical role in his Stage 1 endorsement of PL. Without commitment to Km1, Mendeleev had no procedure for producing values for atomic weights that he felt were reliable. This commitment was an important reason for his endorsement of PL.

How widely shared was commitment to Km1 by the time Stage 1 commences in 1869? British chemists had been exposed to similar ideas as of 1855 through the translation of Laurent's book and the majority of leading chemists in Great Britain had adopted them. Lothar Meyer had attended the Conference and published a textbook in 1864 that "won over a large majority of the German chemists to the new atomic weights." (de Milt 1951: 425) So there is evidence of a considerable degree of commitment to Km1 as of 1869. But there is further evidence that Km1 was not universally accepted by chemists of the period. To begin with, the 1860 Conference itself did not produce a consensus on this point (Sambursky 1971: 8). Mendeleev did claim in his 1889 Faraday Lecture, regarding the years following publication of his 1869 paper, "the ideas of Cannizzaro proved, after a few years, to be the only ones which could stand criticism." (Mendeleev 1889: 636) Holden (1984: 20) cites the publication of Mendeleev's 1869 textbook as one that spread Cannizzaro's methods throughout the scientific community in subsequent years, suggesting that commitment to Km1 was by no means universal at the onset of Stage 1. However, French chemists, under the influence of Berzelius, proved particularly resistant to the new atomic weights over the next twenty years (de Milt 1951: 425). Commitment to Km1 during this period is extensive, but there were dissenters and those who were yet to be convinced. Let us turn to the second of Mendeleev's background beliefs, Km2:

3.7.2 *Km2:* The domain of the elements forms some kind of systematic unity

The periodic law showed that the elements were regulated by a single unifying pattern. Like Newtonian physics and Darwinian biology, the periodic law showed that many apparently disparate phenomena could be given a unified explanation. Mendeleev himself put the point quite eloquently:

> Kant said that there are in the world 'two things we never cease to call for admiration and reverence of man: the moral law within ourselves and the stellar sky above us.' But when we turn our thoughts towards the nature of the elements and the periodic law we must add a third subject, namely, 'the nature of the elementary individuals which we discover everywhere around us'. Without them the stellar sky itself is inconceivable; and in the atoms we see at once their peculiar individualities, the infinite multiplicity of the individuals and the submission of their seeming freedom to the general harmony of nature. (1889: 642–643)

But the general background belief, which I deem Km2, that the elements form some kind of systematic unity was one to which Mendeleev was committed before the onset of Stage 1, as "it is apparent that he acquired a much greater confidence in the feasibility of systematizing extent chemical knowledge than almost any of his contemporaries." (Kultgen 1958: 177) Commitment to the idea of basic elemental unity was "one of the important factors which led Mendeleev to the periodic law." (Bensaude-Vincent 1986: 7) Mendeleev was therefore quicker than others to perceive a single unifying pattern when he ultimately encountered suggestive evidence.

An analogy might help to make this point: suppose Sherri were presented with a sequence of photographs, of varying clarity and quality, of individual puzzle pieces. Suppose Sherri studies the photographs and endorses theory T: the pieces fit together to form a picture of Winston Churchill. She displays how she believes they fit together – she assembles the various pieces, some of which seem to fit in her picture precisely as she says they should. But she allows that because of the uneven photographic quality, a fair number of the pieces have to be reshaped to make the whole picture come out looking as she says it should. Critics might object, pointing to the looseness of fit as evidence against T. But suppose Sherri is committed to the background belief that the pieces fit together to form a coherent picture. Her background belief makes her more ready to believe that she has discovered the hidden picture and thus more ready to revise pieces of the evidence that she needs to make her theory work. But evaluators of T who were not committed to this belief (who believed that the pieces may well not fit together into a coherent picture) would be much more skeptical about T. Likewise, the belief that the chemical elements did form a unified structure bolstered Mendeleev's confidence in his version of PL, and skepticism about a general structure generated skepticism among evaluators during Stage 1 (as we will see).

What was the basis for Mendeleev's commitment to Km2, apart from his detailed knowledge of the elements and their properties? Kultgen writes that Mendeleev's confidence that he would find a theory that systematized the elements proceeded from his overall conception of science itself:

Apparently Mendeleev conceived the scientific enterprise to *be* a search for system. Accept the enterprise, and you are obliged to accept the possibility of the system. Of course, the question may be raised, Why have confidence in the enterprise? Mendeleev apparently did not feel the need to raise this question. He firmly believed in the future of science, and it is quite possible that he held this faith for a-logical reasons. (1958: 182)

Kultgen goes on to cite as possible examples of such a-logical reasons Mendeleev's devotion to his mother, who encouraged a reverent view of the fertility of science in her dying words to him, and the fact that Mendeleev's greatest intellectual gifts were in his ability to systematize information. Kultgen muses, "One is happy to assume the possibility of an activity which engages one's best talents." (1958: 182 n.13) He implies that an early commitment to something like what I call Km2 was not grounded on evidence, but on extra-scientific reasons.

I do not agree. There is evidence that suggests that Mendeleev's antecedent commitment to Km2 derived from an independently grounded metaphysical view about the unity of nature. In the Faraday Lecture he states that in response to the legitimate scientific tendency to find unifying explanations for disparate phenomena

... natural science has discovered throughout the universe a unity of plan, a unity of forces, and a unity of matter, and the convincing conclusions of modern science compel everyone to admit these kinds of unity. But while we admit unity in many things, we nonetheless must also explain the individuality and the apparent diversity which we cannot fail to trace everywhere. It has been said of old, "Give a fulcrum, and it will become easy to displace the earth." So also we must say, "Give anything that is individualized, and the apparent diversity will be easily understood." Otherwise, how could unity result in a multitude? (1889: 645)

Thus Mendeleev believed in the claim that nature was in general a unified structure, and that this claim had been shown by antecedent successful scientific ventures. He thus had been, as of 1869, in a position to induce from these earlier successes to the domain of elements and conclude that they too probably admitted of a general unified structure.

How widespread was belief in Km2 during Stage 1? The story is a rather complex one. As of the early nineteenth century, there was no real evidence for any kind of order among the elements (Strathern 2000: 256). Chemistry was at this time little more than a mass of empirical facts about substances that bore no particular relation to one another. Describing the state of chemistry at this time in comparison to the state of physics, Bensaude-Vincent writes:

Clear and deep physical investigations of matter [in physics] were fundamentally opposed to obscure, intricate and superficial chemical knowledge [in the eighteenth century]. Was this overwhelming statement quite obsolete in the nineteenth century? As chemistry dealt with individual properties of bodies and with their specific mutual reactions, it still naturally lacked the clear generality of physics and seemed to be condemned to empiricism. Memorizing the names of hundreds of chemical substances, learning their reactions and transformations, such was the challenge for a student of chemistry at the end of the 18th century. (1986: 4)

Bensaude-Vincent points out that, as of the 1860s, while the challenges facing the student had been alleviated by the introduction of Lavoisier's greatly simplified nomenclature, the growing number of 'simple substances' (substances which could not be decomposed into components) made learning the basics of chemistry very complicated for beginners. She writes "The endless growing population of elements seemed to condemn chemistry to something like a natural history of chemical substances. Once again, chemistry was far from the clarity of physics because of the unavoidable pluralism of simple substances." (1986: 5) As of the mid-nineteenth century, that is, the domain of the elements, far from forming a simple and unified structure, was a rather disunified mess. Mendeleev thus noted in his Faraday Lecture, concerning the 1860s, that the "idea of seeking for a relation between the atomic weights of all the elements was foreign to the ideas then current ..." (1889: 638, printed in Knight 1970: 326). The scientist Marcellin Berthelot was to discourage enthusiasm for Mendeleev's PL as late as 1888, in his book on alchemy, by describing it as one of those "theories that spring from the desire for unity and causality inherent in the human mind ... But such is the seduction exercised by these dreams, that it is necessary to guard against seeing in them the fundamental laws of our science and the basis of its facts, on pain of falling again into a mystic enthusiasm parallel to that of the alchemist" (quoted in Venable 1896: 113).

However, there was a discernible trend toward recognition of pattern among the elements at this time as well. Perhaps the earliest clearly identifiable precursor to the Periodic Table was Dobereiner, who first identified a relationship between atomic weights (more accurately, equivalent weights) of elements (and compounds of elements) with analogous properties in 1817. He noted that various chemically similar elements could be arranged in 'triads' – one element or compound in the triad had an atomic weight that was the average of the weight of the other two. Various chemists extended Dobereiner's program in the mid-nineteenth century, including Gmelin, Pettenkofer, Dumas, Kremers, Gladstone, Lenssen, and Cooke (van Spronson 1969: 63–95, and Scerri 2007 Ch. 2). These chemists clearly discerned to differing degrees systematic relationships between elements – in particular, those that are noticeably similar to each other – involving arithmetic patterns among atomic weights. So it is the case that there is a growing realization that the elements are subsumable into some kind of system by mid-century. On the other hand, there is a difference between this kind of view and the actual content of Km2. The chemists mentioned above are described by Bensaude-Vincent as 'classifiers': they proposed in various ways to group elements into families of elements based on their chemical and physical similarities. But they did not discern a single unifying pattern that covered all the elements – thus they did not in any serious sense embrace Km2 – which posits a systematic unity among the elements. Commitment to Km2 can be discerned in those most often considered as co-discoverers of PL with Mendeleev: Newlands (in his 'law of octaves'), de Chancourtois, Meyer and Odling (cf. Cassebaum and Kauffman 1971).

Another point worth discussing that bears on Stage 1 commitment to Km2 concerns Prout's hypothesis. In its original form, proposed by the London physician William Prout in 1816, it claimed that, since it appeared that the atomic weights of many elements were integral multiples of the atomic weight of hydrogen, hydrogen was the 'primary matter' of which all other elements were composed. Later more accurate determinations of atomic weights brought in non-integral values for atomic weights, but Prout's hypothesis survived into the middle of the nineteenth century thanks to the postulation that the primary matter was not hydrogen but an unknown element which should be a quarter or a half of hydrogen's atomic weight. Although some chemists (like Mendeleev himself) rejected Prout's hypothesis in any form as mere speculation, serious interest in this hypothesis "was not an eccentric and unpopular movement, but involved many of the great names of nineteenth-century chemistry . . ." (Farrar 1965: 291). The vindication of Prout's hypothesis would represent a vindication of Km2 – for by providing a single ur-element from which all other elements could be composed it provided a unified account of the domain of the elements. Nearly all Mendeleev's precursors (with the exception of Cooke, Odlings and Newlands) were advocates of Prout's hypothesis (Bensaude-Vincent 1986: 6). In this respect, there was a fair amount of enthusiasm in mid-century for Km2, more precisely, for a particular conception of what elemental unity amounted to. Such enthusiasm did not translate directly into enthusiasm for PL, as it advocated a different (though possibly related) form of unity altogether. We turn now to the third and final example I will offer of Mendeleev's background beliefs.

3.7.3 *Km3:* Theoretical inference can be truth conducive in chemistry

As I will document below, some of the skepticism PL faced from the chemical community was a result of the 'theoretical' nature of PL. As of the mid-nineteenth century, chemistry was characterized to a significant extent by a positivistic conception of method that sought empirical knowledge – facts established by experiment – and spurned theory. As we will see, the success of Mendeleev's predictions about the properties of various then undiscovered elements was astonishing to many chemists who had strongly believed that the only way to know of the existence or the properties of an element was to perform experiments that revealed such information. Mendeleev's success demonstrated that experimentalism was not the only path to knowledge – the construction of theory was another such path (such construction must itself be grounded, of course, on observation and experiment). But all of this raises one of the classic questions in the philosophy of science: what exactly is a theory? One popular answer to this question is that a theory is a statement that posits unobservable entities or properties. But the skepticism that Mendeleev faced that is at issue here did not pertain to any postulation of unobservable entities – for he posited none of a sort different than those accepted by many others. Rather, the skepticism he faced was attributable to the fact that he endorsed a hypothesis which had posited new types of entities that had not at that point been observed. It was Mendeleev's inference to PL from his accommodated data that was perceived by some as suspiciously 'theoretical' – for it seemed absurd that information about the existence and nature of something like an element could be known in advance of direct empirical substantiation.

I suggest therefore that for present purposes a 'theory' is any hypothesis which posits entities of a sort not heretofore directly observed (notwithstanding the vagueness of the concept of 'direct observation' which I propose to tolerate). This conception subsumes the usual conception of a theory as a hypothesis that postulates unobservable entities but covers other sorts of hypotheses as well. Theoretical inference then is any inference from some established premises to a theory thus defined. Km3 asserts that such theoretical inference can be reliable, viz., it is in some cases truth-conducive. My position is that Mendeleev was strongly committed to Km3 prior to Stage 1 – and indeed such a commitment is necessary to explain his strong commitment to PL prior to the direct observation of the various elements and properties he predicted (for Mendeleev, "with his very detailed predictions of undiscovered elements, had no doubt about the importance of his discovery and staked his reputation on it." (Knight 1970: 11))

How widely accepted was Km3 during Stage 1? There is evidence of a considerable skepticism about it among chemists during Stage 1 and during the decades before it. Commenting on the state of chemistry in the early nineteenth century, Strathern writes that chemistry "had achieved its scientific status largely through experiment, and such theoretical thinking [about any fundamental pattern among the elements] was viewed at best as mere speculation." (2000: 256) Important pre-cursors of the periodic law were developed by Newlands and Odling, both of whom presented early versions of PL to the Chemical Society of London in the mid-1860s, which refused to publish their results in their Society's journal because of a policy against publishing purely theoretical work – a policy based on that Society's terror of what it regarded as "mere speculation" (Knight 1970: 11). Newlands, having presented his law of octaves to this Society in 1864, was sarcastically queried whether he had considered the merits of placing the elements in alphabetical order as well as sequencing them by their atomic weight. Much later Ernest Rutherford reflected on the lack of interest shown by the chemical community in Mendeleev's theory during Stage 1, and noted that many chemists had a strongly empirical approach to their field, as chemists "were more interested in adding to the chemical facts than in speculating on relationships between them." (Holden 1984: 23) Rutherford implies, it seems to me, a pervasive lack of confidence in Km3 among the chemists of this period. The most prominent Swedish chemist of the nineteenth century, Berzelius, adopted an explicitly instrumentalist view of theory in chemistry, and Swedish textbooks of this period strongly emphasized the presentation of empirical facts and were explicitly atheoretical (Lundgren 2000). Nineteenth-century chemistry textbooks in France were similar in nature, particularly in early and mid-century, presenting an image of chemistry that was, at its current state, constituted by empirical knowledge of individual substances, though such textbooks implied that this was a temporary state that would one day give way to a theoretical science (Bensaude-Vincent 2000: 285). The controversy over Km3 was felt too at the 1860 Karlsruhe Conference – Mendeleev wrote in a letter to a former teacher that one of the proposed topics of discussion was whether "in the present state of science, we should consider the reasons for chemical effects" (printed in de Milt 1951: 422). Clearly, the chemists of Stage 1 did not universally accept Km3 – though Mendeleev himself was committed to it at least to the extent necessary to underwrite his confidence in PL.

I have argued that Mendeleev's high confidence in PL was attributable in part to his commitment to three background beliefs about which he was very confident. Together with the accommodated evidence about atomic weight values and chemical and physical properties of elements, these background beliefs strongly supported PL. Furthermore, widespread skepticism about PL during Stage 1 is easy to understand, given that there was notable disagreement throughout the chemical community about these background beliefs (particularly Km2 and Km3).

It is time, at last, to turn to the question whether and in what sense the evaluators of PL were predictivists. As noted, predictivism can take four forms. It is my position that there were probably evaluators in the relevant scientific community who correspond to each of these forms, though in some cases it is easier to point to specific examples of characteristic evaluators than in others. We will deal with each in turn.

3.8 PREDICTIVISM AND THE PERIODIC LAW

3.8.1 *Thin virtuous predictivism*

Thin predictivism is a species of weak predictivism which holds for individualist evaluators who deny any actual epistemic significance to predictive acts but hold that such acts are linked to some other factor that increases theory probability. Thin virtuous predictivism holds that this other factor is (1) the existence of reasons – now described as background beliefs – that prompted the initial endorsement of the theory, and (2) the fact that the background beliefs are more strongly confirmed by the demonstrated success of the predictions than by a similar amount of accommodated evidence.

Individualist evaluators who surveyed Mendeleev's total Stage 1 evidence regarding PL would thus identify his three background beliefs (and perhaps others), together with the specific information about the known elements, as underwriting his strong endorsement of PL – this identification constitutes the 'per se' side of thin virtuous predictivism. What remains to be considered is the 'success' side of Stage 2. In what follows I will consider the disposition of Mendeleev's successful predictions – as opposed to his accommodations – to confirm his various background beliefs for individualist evaluators. In the case of two of these beliefs, I attribute a virtuous predictivism to these evaluators, and in the other case I do not (except in a qualified sense).

3.8.1.1 *The confirmation of Km1*

Km1 asserts that the methods of atomic weight estimation endorsed by Cannizaro at the 1860 Karlsruhe Conference are sound and that many earlier estimations of atomic weight were inaccurate. The sound methods included the determination of vapor densities used earlier by Gerhardt. Km1 was obviously critical for Mendeleev's endorsement of PL because it was by assuming Km1 that he took himself to have accurate estimations of many atomic weight values as of 1869. Now I have heretofore described Mendeleev's accommodated evidence as including these very atomic weight estimations, along with many other claims about physical properties like boiling point, melting point, density, and chemical properties involving dispositions to react with other elements to form particular compounds. Obviously on this description his accommodated evidence suffered from a serious inability to confirm Km1: the various estimations of atomic weights Mendeleev used as accommodated evidence could not confirm Km1, because one needed to assume Km1 in order to have confidence in the estimations themselves! This entails that any evaluator skeptical about Km1 (and I have argued above that there were at least some of these) would be skeptical about some of the very evidence we are describing Mendeleev as 'accommodating.' But this is problematic – for the basic idea of accommodated evidence is that it is evidence that is counted as 'actual' by endorser and evaluator alike (though it is an open question how much probative weight the evidence has at this point). My proposal is to reconstrue the identity of accommodated evidence – for the purpose of considering the confirmation of Km1 – as including not the estimated values of the 'known' atomic weights, but as including the laboratory data that was used in computing these weights by those who assumed Cannizaro's methods of atomic weight computation. I will refer to this data as the 'atomic weight data.' These data were the result of 'direct observation' in the laboratory and can be counted as actual by supporters and detractors of Km1. Mendeleev's accommodated evidence, now, consists of the established atomic weight data along with the established chemical and physical properties of the known elements (but not the computed values of their atomic weights). Mendeleev's predictions, it seems to me, should now be described as predictions about the atomic weight data to be obtained for such elements along with their other properties.

Now let us pose again the question whether predictivism held true for a hypothetical individualist evaluator – who is at least somewhat skeptical about Km1 – in the course of her evaluation of Km1. Would Mendeleev's successfully predicted atomic weight data offer a qualitatively greater confirmation of Km1 than his accommodated atomic weight data? I believe the answer is a clear 'yes,' at least for some easily imaginable individualist evaluators of PL. One such evaluator could reason to himself as follows: skeptical as I am about Km1, I cannot take Mendeleev's accommodated atomic weight data as entailing any particular conclusions about actual atomic weights, thus I cannot take the data as supporting Mendeleev's version of PL. However, it must be conceded that his accommodated atomic weight data *in toto* seems to be characterized by a particular pattern (one such that, assuming Km1, atomic weights are entailed which, in conjunction with other known chemical and physical facts, constitute PL). Mendeleev's predictions about the discovery of heretofore undiscovered elements, and about the eventual correction of currently accepted values for atomic weights, are extrapolations of this same pattern of atomic weight data. The probability of the extrapolation being confirmed, if Km1 is true, is reasonably high, for if Km1 is true then Mendeleev's atomic weight values are correct, and thus they serve to confirm PL fairly strongly. But the probability of the extrapolation being confirmed if Km1 is false is quite low – for if Km1 is false then no conclusions about atomic weights are entailed by Mendeleev's data, and PL is entirely unsupported, leaving us with no reason to expect his predictions to be confirmed. Thus, if we let N = Mendeleev's predicted data, it follows that the ratio of $p(N/Km1)/p(N/\sim Km1)$ is quite high – and N offers much confirmation of Km1 (unlike the accommodated atomic weight data).

To illustrate my position another way I offer an analogy: in John Locke's *Essay Concerning Human Understanding* he provides various arguments designed to support the claim – contra the skepticism of Descartes' First Meditation – that our sensations are accurate pictures of the external world.

These arguments appeal to the fact that our various sensations occur in reliable and predictable patterns. The perseverance of such patterns over time was evidence, Locke argued, that our senses are accurate sources of information about the external world – for what else but their accuracy would explain such as astonishing perseverance?[9] (Locke (1995: 539) Locke's analysis suggests a similar bifurcation of sensory evidence into accommodated evidence – early sensations which serve to suggest a pattern or regularity in our sense data – and predicted evidence – subsequent sensations that confirm the implications of that apparent pattern or regularity.

Locke argues such subsequent sensations confirm the veracity of sensation itself. Say Ks is the background belief that sensation is reliable. He apparently reasons as follows: early sensations by themselves provide no indication that Ks is true, but early sensations combine with Ks to entail a general account of what the world is like, which in turn entails predictions about subsequent sensations. These subsequent sensations, if obtained, thus confirm Ks. A similar argument, I propose above, works to defend the special confirming power of some of Mendeleev's predictions vis à vis PL: his accommodated data provide no indication that Km_I is true, but accommodated data combine with Km_I to reasonably support PL, which in turn entails novel predictions N. N, once confirmed, confirms Km_I, for the accommodated data in effect makes N a (non-strictly entailed) consequence of Km_I.

The Lockean story just told involves a bifurcation of sensory evidence into accommodated and predicted evidence but gives no details about when, where or how this bifurcation should be effected. In truth, there is a similar indeterminacy in the story of Mendeleev's evidence. An evaluator might consider all of Mendeleev's accommodated evidence as collectively sufficient to suggest a 'pattern' of the sort described which, combined with Mendeleev's background beliefs, is sufficient to suggest (read: endorse) PL – but he might regard some proper subset of this evidence as by itself sufficient, combined with Mendeleev's background beliefs, to suggest PL. If the latter is true, he will regard some of Mendeleev's officially

---

[9] Of course, there are skeptical hypotheses such as Descartes' evil genius hypothesis that are suitably rigged so as to be empirically indistinguishable from the hypothesis that the senses are reliable, and Locke's argument does not succeed against them. Locke's argument nonetheless has some force in common sense contexts where such bizarre competing hypotheses are not at stake. One might worry, analogously, that novel confirmation of Km_I cannot work to confirm Km_I given the existence of hypotheses which are empirically equivalent to Km_I with the same novel consequences. I show in Chapter 5, however, that such anti-realist arguments ultimately fail to repudiate the reality of predictivism.

'accommodated' evidence as actually predicted – and come that much more quickly to a high opinion of PL. My position is simply that there will be, for the evaluator in question, some bifurcation of evidence into accommodated and predicted evidence, and that the latter has a qualitatively greater ability to confirm Km1.

Let us now take a careful look at the first major novel success: Lecoq de Boisbaudran's discovery of gallium. Mendeleev had predicted the eventual discovery of an element he called eka-aluminum – whose atomic weight Mendeleev estimated to be about 68. His estimation was based on a computation of the mean of the atomic weights of the elements its neighboring elements in the Periodic Table. De Boisbaudran's discovery employed spectroscopy, a form of chemical analysis that de Boisbaudran himself had done much to develop. He had asserted as early as 1868 that the wave lengths produced by a spectroscopic analysis of an element provided an accurate measure of that element's atomic weight – in August of 1875 de Boisbaudran used this very method to detect the existence of an element whose atomic weight he estimated to be 69.86, which he named 'gallium' (Spring 1975: 135). The chemical and physical properties of gallium also coincided with those predicted by Mendeleev to an astonishing degree. This novel success could thus be thought of as having multiple aspects, corresponding to all the chemical and physical properties predicted and subsequently confirmed. Our present question is whether this novel success – in toto – offered any special confirmation of Km1. But this question seems to have an easy answer: of course it did. The assumption of Km1 had been one key premise in Mendeleev's endorsement of PL – this critical confirmation of PL thus redounded to the credit of Km1. There is no circularity in this claim to confirmation, but this is in part because de Boisbaudran's estimation of gallium's atomic weight was not based on Cannizaro's methods, but on the newer methods of spectroscopic analysis, which had won considerable credibility in the years since Cannizaro's 1860 address. Thus this novel success offered confirmation of Km1 because it was itself credibly determined without having to assume the truth of Km1 – unlike the values of other established atomic weights.

This analysis invites the following reply: though the discovery of gallium clearly offered a form of confirmation for Km1 fundamentally greater than could be offered by the accommodated evidence, this had nothing, really, to do with the novelty of the discovery, but simply reflected the fact that the discovery was based on a method of atomic weight estimation that was entirely independent of Cannizaro's methods. Its novelty was, from the standpoint of the confirmation of Km1, fortuitous.

I believe it is worth pausing for a moment to pose the following hypothetical question: what would have been the case if the discovery of gallium
had, contrary to historical fact, been based on the sort of methods endorsed
by Km1? Would it thus have been impotent to offer confirmation of
Km1? My position is that the answer is 'no,' at least for some imaginable
evaluators. My reasons are given in the account provided above: at least for
some imaginable evaluators, such a discovery could have provided important evidence for Km1 insofar as it would represent a confirmation of a
pattern suggested by the accommodated atomic weight data in tandem
with Km1.

I have argued that the discovery of gallium constituted a confirmation of
Km1 qualitatively different from that provided by accommodated evidence
in point of historical fact – and (more importantly for our philosophical
purpose) that it would have provided such confirmation in a hypothetical
scenario in which the discovery was based on the same methods used in the
acquisition of Mendeleev's accommodated evidence. All of this will hold
for evaluators who accord no epistemic significance to Mendeleev's act of
endorsement who furthermore take Mendeleev's strong endorsement of PL
as based on a correct assessment of the implications of his background
beliefs together with the available accommodated evidence. This reveals the
possibility of a thin predictivism for individualist virtue-ascribing evaluators in the confirmation of Km1.

3.8.1.2 The confirmation of Km2

Km2 asserts that the domain of elements forms some kind of systematic
unity. I argued at some length above that Mendeleev's early commitment
to Km2 was an important factor that led him to endorse the law – and that
there was a fair amount of skepticism about Km2 during Stage 1 across the
chemical community. It is easy to agree that the astonishing success of
Mendeleev's various predictions brought important confirmation to Km2
for evaluators who had previously been skeptical about it. But to agree to
this is not necessarily to agree to any kind of predictivism – but simply to
the claim that the success of Mendeleev's predictions left Km2 more
strongly confirmed than it had been given just the old evidence. What is
less obvious is whether the confirmation of Km2 by the novel successes was
any greater than the confirmation provided by a comparable number of
pieces of accommodated evidence.

We must address again the question of how to characterize the accommodated evidence. In the previous section it turned out that the accommodated evidence was construed as including not Mendeleev's values for

the various atomic weights of the known elements but as the 'atomic weight data' that was obtained in laboratories that sought to measure such weights by assuming Km1. This was done to discern a thin virtuous predictivism in the case of Km1. Now in the present case of Km2 evaluation we might imagine evaluators of two types: (1) an evaluator who assumes, as early as Stage 1, that Km1 is true and (2) an evaluator who does not accept Km1 during Stage 1 but finds it strongly confirmed in Stage 2. Whether we imagine a 1-type or a 2-type evaluator could determine how we characterize the accommodated evidence vis à vis Km2 as well. A 1-type evaluator would regard Mendeleev's accommodated evidence as including the values of the atomic weights of the known elements used by Mendeleev in endorsing PL. Since he accepted Km1 as a background belief, he would regard such values as known, and thus as accommodated evidence. I see no reason to think that the confirmation of Km2 would be characterized by predictivism for the 1-type evaluator. Both the accommodated and the predicted atomic weights find their place in the unified pattern that characterizes PL, and both seem to point to its unity with equal force. However, matters are different for the 2-type evaluator, for whom the accommodated evidence would include the atomic weight data that underwrote Mendeleev's judgments about the atomic weights of the known elements. During Stage 1 this accommodated data was shown to be powerless to confirm Km1 for the 2-type evaluator (cf. the previous section) – but then it must have been powerless to confirm Km2 as well. For without much reason to think that Km1 is true, there was no reason to accept the values of the atomic weights accepted by Mendeleev, and thus no reason to view Km2 as supported by the values of those weights as estimated by him. But with the coming of Stage 2 confirmation of Km1, Mendeleev's estimated atomic weight values were confirmed, and thus the apparent unity of the elements as discerned by Mendeleev was confirmed. Thus the predictivism that characterized the evaluation of Km1 would trickle down to Km2 for the 2-type evaluator.

3.8.1.3 The confirmation of Km3

Km3 asserts that theoretical inference in chemistry is truth conducive. I proposed above that a 'theory' in our present context is any hypothesis which postulates heretofore unobserved types of entities. Let us consider an evaluator who is somewhat skeptical about Km3 (of whom there were many, as shown above). Such an evaluator is of course destined to be skeptical about PL, given its novel content. Moreover, it seems to be the case that this sort of skepticism can withstand any amount of accommodated evidence provided PL retains some content not directly established by observation. But what would a Km3 skeptic make of Mendeleev's various novel confirmations? If such a skeptic were sufficiently dogmatic in her Km3 skepticism, she might persist in regarding all the subsequent novel successes as mere coincidence. But a less dogmatic skeptic would see in the novel consequences evidence of the truth of Km3, and thus presumably evidence of the truth of Km3. (The difference between a dogmatic and less dogmatic skeptic would consist – in terms of the Bayesian model developed in this chapter – of a difference in prior probability assigned to Km3.) Thus we find the basis for a thin virtuous predictivism for individualist evaluators initially skeptical about Km3.

Is there any actual historical evidence for a dramatic shift toward Km3 as the result of Mendeleev's novel confirmations? I believe there is. Brush describes one consequence of the success of Mendeleev's predictions this way:

For some chemists, the success of Mendeleev's predictions meant that their science had at last attained the respected status long enjoyed by astronomy and physics. This boost in prestige had been foreshadowed in the words of F. W. Clarke's 1878 address to the AAAS ... Astronomy's spectacular 1846 triumph in predicting and discovering a new planet was still a vivid memory for the older generation and a well known historical event for the younger: the cases of gallium, scandium, and germanium were compared to that of Neptune as demonstrations of the capacity of mature exact science. (Brush 1996: 616)

F. W. Clarke, an American geochemist, had in the address mentioned above (which omitted any reference to PL) claimed that

... one of the main objects of science is to render prevision possible. The more thoroughly our knowledge is coordinated, the better we are able to predict the nearer discoveries of the future and to see what lines of research will be most fruitful ... To-day, notwithstanding its brilliant achievements in the past, chemistry is an inexact science ... in those principles which render foresight possible, chemistry is poor and vague. We may guess the existence of some undiscovered compound, but until we have prepared it, what can we tell of its properties? (quoted in Brush 1996: 611)

I take it to be clear that those "principles which render foresight possible" include theories that are sufficiently well confirmed to enable confidence in their novel consequences. For chemistry to be poor in such principles is for it to lack theories that merit such confidence – hence the widespread skepticism about Km3. Six years later, in a well-known textbook, Clark praised the periodic law and said that the predictions of gallium and scandium were "now regarded as among the most remarkable achievements of modern science" (Clarke 1884: 182, quoted in Brush 1996: 611–612). Likewise C. S. Peirce wrote that PL, in the light of its predictions, "may be ranked as second to the research of Kepler into the motions of Mars" (Peirce 1897: 424).

Such comments indicate the existence of evaluators who take the epistemic significance of Mendeleev's successful predictions to apply not simply to Mendeleev's theory but, in some sense, to the entire field of chemistry. The spectacular confirmation of PL had wider ramifications – and testified to the validity of theoretical inference in chemistry in general. All of this reveals a virtuous predictivism in the scientific community's assessment of Km3 that was accessible to individualist virtue-ascribing evaluators.

When Lecoq de Boisbaudran initially reported his discovery of gallium, he described it as having many physical and chemical qualities that coincided strongly with values predicted by Mendeleev. However, the value of its density differed somewhat from the predicted value, a point that Mendeleev observed in a note entitled "Note on the Occasion of the Discovery of Gallium." As a result of this observation de Boisbaudran recomputed its density, only to find Mendeleev's value corroborated. Consequently "the scientific world was astounded to note that Mendeleev, the theorist, had seen the properties of a new element more clearly than the chemist who had empirically discovered it. From this time, too, Mendeleev's work came to be more widely known . . ." (Gillespie 1974: 290; cf. also Brush 1996: 603–604) Likewise Holden writes that "Scientists were extremely impressed that Mendeleev could know more about a new element from theory than the discoverer of that element could from his experiment" (1984: 29). The implication is that the epistemic significance of this novel success was partly in its power to confirm Km3.

I have argued for thin virtuous predictivism in the scientific community's assessment of Mendeleev's background beliefs Km1 and Km3, but I have found no basis for such predictivism in the case of Km2 (though even here there is room for a derivative predictivism.) My position is that one important explanation of the fact that the scientific community came to embrace PL upon the arrival of Stage 2 is that some members of this community were epistemic individualist and virtue-ascribing evaluators. We turn now to the case of the thin unvirtue-ascribing evaluator.

3.8.2 *Thin unvirtuous predictivism*

To review: the thin unvirtue-ascribing predictivist is an individualist who accuses accommodators of a certain kind of incoherence – the error of misrepresenting the probability of some theory given some body of accommodated evidence, typically by giving that probability an estimated value that is too high, given the accommodator's logical dishonesty or logical ineptitude (cf. 3.2). In the usual case, the endorser is someone who the evaluator regards as overly committed to the theory at issue, say because he is or was the first to propose it or because he publicly endorsed it. Thus the endorser is prone to be mistaken (at least in his public acts of endorsement) in how well the total accommodated evidence (plus his background beliefs) supports the theory. A thin unvirtue-ascribing evaluator of Mendeleev's PL, then, would be someone who accused Mendeleev of judging PL (during Stage 1) as more strongly supported by the total accommodated evidence and background belief than he was entitled to be, given his other commitments. Such an evaluator, however, should be impressed by the PL's novel successes, for a pre-success endorsement leaves the endorser no opportunity to cook his endorsement in an unvirtuous way.

One skeptical evaluator who certainly did accuse Mendeleev of logical error was the French chemist Marcellin Berthelot (1827–1907). Berthelot argued that the "successes of the law should be credited to the previously recognized regularities of the atomic weights in families of elements, along with the 'convenient trick' of creating a net so fine grained ... that no new body, whatever it may be, can fall outside the meshes of the net." (quoted in Brush 1996: 613) Berthelot clearly regarded Mendeleev as failing to understand that the evidence he cited did not support PL as strongly as Mendeleev claimed – for he implied that Mendeleev failed to notice that the patterns that constituted PL were structured in such a way as to virtually guarantee a place for any element that might happen to exist. Thus Berthelot would appear to qualify as a thin unvirtue-ascribing evaluator. However, the matter is complicated somewhat by the fact that Berthelot's criticism came well after Stage 1 was over – in his 1885 book on the origins of alchemy! Berthelot's skepticism thus had survived the discovery of gallium and scandium and the corroboration of Mendeleev's predictions about the actual atomic weights of uranium and beryllium. This skepticism appears to make sense given the nature of Berthelot's criticism – which is that the known properties of any element (whether novelly predicted or not) could be made to fall neatly into place in the Periodic Table. However, though I have no direct substantiation of the point, there is reason to think that his skepticism eventually was vanquished by yet further novel successes. Commenting on the 1885 criticism, Brush notes that Berthelot's timing was bad. The following year brought Winkler's discovery of germanium, and no further criticism from Berthelot was forthcoming. By the late 1880s the periodic law was widely accepted. (Brush 1996: 614) It is at least plausible to conjecture that Berthelot was, in the last analysis, a thin unvirtue-ascribing

predictivist – as his criticism of PL was apparently ultimately silenced by yet
another novel success.

3.8.3 Tempered predictivism: virtuous and unvirtuous

We return now to the perspective of a pluralist evaluator – one who does
regard the posted probabilities of other agents as having epistemic import.
In this case, we consider whether there were evaluators who took
Mendeleev's act of endorsement, viz., his predictive act, as indicative of
the possession of reasons that the evaluator does not take herself to possess,
or at least not fully possess – and who took the success of Mendeleev's novel
predictions to constitute special evidence for the truth of those reasons. In
the previous chapter a typology of pluralist evaluators was developed –
pluralist evaluators included non-experts, interdisciplinary experts, imperfect experts, humble experts, and reflective experts. I will argue that the
complete set of Mendeleev's evaluators probably included each of these. The
evaluation of PL – during Stage 1 and at least in early Stage 2 – took place in
a highly pluralistic environment. But I argued in Chapter 2 that if an
epistemic environment is pluralistic, it is probably destined to be a predictivist environment. Thus to the extent that it is possible to show that the
environment was pluralistic, the conclusion that it was predictivistic is
strongly supported. The well-documented dramatic shift in scientific opinion toward Mendeleev's theory during Stage 2 can be thereby explained.

There are, of course, two forms of tempered predictivism: virtuous and
unvirtuous. The virtue-ascribing evaluator regards the endorser as logically
impeccable while the unvirtue-ascribing evaluator views the endorser as
logically suspect. As has been explained, the two forms are associated with
different bases for predictivism. The pluralist evaluator generally does not
know (or fully know) what the epistemic basis for the endorser's posted
probability is – this means that the evaluator may have no direct knowledge
of whether the endorser's logic is correct or not. A pluralist evaluator thus
could be agnostic on the question of the endorser's logical trustworthiness.
But, assuming the evaluator regards it as at least possible that the endorser
is not logically trustworthy, the evaluator will be prone to value novel
confirmation more strongly for reasons associated with unvirtue-ascribing
evaluation: successful prediction will transcend a modest skepticism about
accommodation as well as a severe skepticism. But the same evaluator
could value novel confirmation for reasons associated with virtuous predictivism as well: novel success confirms the truth of the endorser's reasons
for the endorsement. Thus the predictivism of pluralist evaluators can rest

on both virtuous and unvirtuous bases, or on their disjunction. In what follows we will collapse the distinction between virtue-ascribing and unvirtue-ascribing evaluation and consider a single category of pluralist evaluator and the tempered predictivism that will, arguably, characterize her theory evaluation (theory evaluation by 'agnostic evaluators' – those who do not know whether endorsers are virtuous or unvirtuous – is discussed in more detail in Chapter 6).

If evaluators of this period were pluralistic, we would expect that they would be impressed, first of all, by the enormous confidence with which Mendeleev endorsed his theory, and thus his predictions. But this seems to have been the case. The famous English chemist Thomas Thorpe wrote in 1907 that "The amazing accuracy of Mendeleev's predictions and the obvious confidence with which they were made stunned the whole scientific world and guaranteed for them and their inferences esteem and recognition, the like of which it was difficult to expect in another case." (quoted in Solov'ev 1984: 1069). I propose that Thorpe's reference to the epistemic significance of Mendeleev's confidence vis a vis PL is a clear illustration of the power of per se predictivism to influence the evaluations of PL – and of Mendeleev's various background beliefs – in the minds of pluralist evaluators. Thorpe here supports the claim that part of what made PL compelling was that it was very probable for Mendeleev, viz., highly probable on his various background assumptions. His reference to the accuracy of his predictions testifies to the reality of success predictivism for pluralist evaluators.

In my view the most important form of pluralist evaluator in this historical episode is the imperfect evaluator. Imperfect experts were identified in Chapter 2 as experts who have an incomplete grasp of their own specialty. This could consist of their failing to understand the content of all theories that fall within their field, failing to know all the extant data and empirical knowledge collected by colleagues, etc. In Chapter 2 I identified as examples of imperfect experts a large body of British physicists who ultimately served as evaluators of Einstein's General Theory of Relativity. While these scientists knew of the existence of Einstein's theory, they did not make a serious effort to understand and assess the theory until it was strongly endorsed by two prominent British physicists. It seems clear that the British physicists were at this point pluralists who accorded epistemic significance to other scientists' acts of endorsement. What furthermore confirmed Einstein's theory was the highly publicized fact that the theory had generated an impressive novel confirmation that was (apparently) confirmed in 1919 by the Eddington eclipse expedition. This brought considerable attention to the theory – and scientists began to study it in earnest.

It seems to me quite obvious that the story of the evaluation of Mendeleev's PL is strongly similar to the story of General Relativity. Upon the onset of Stage 1, news of the existence of his theory spread quickly – besides the publication of his 1869 and 1871 papers in Russian, a brief abstract of the 1869 paper was published in German in a reputable journal. But those chemists who knew of the theory purely on the basis of the published abstract had of course only a cursory familiarity with the theory's content and very little awareness of the evidence that supported it. It is even unclear how much understanding Russian chemists took away from their reading of Mendeleev's published work. In explaining the failure of chemists to show much interest in PL during Stage 1, it has been observed that Mendeleev's 1871 paper (which contained a more complete version of the table than that given in the 1869 paper, and which appeared in translation in a German journal late in 1871) was too long to read comfortably and contained too much detail (Partington 1964: 896–897). There were, surely, many chemists during Stage 1 who should be described as imperfect experts in the sense that they lacked both a detailed understanding of a theory that had been proposed in their discipline and any real understanding of the various arguments that had been offered on behalf of that theory. There is support for this claim in the body of historical literature that describes the chemical community not as rejecting PL during Stage 1 but as 'neglecting' it – until it began to enjoy novel confirmations. Large numbers of scientists who knew of the theory simply did not give it serious study during Stage 1. Spring (1975: 184) describes Mendeleev's 1869 and 1871 papers as 'neglected,' and Partington holds that it was only after the success of Mendeleev's predictions of missing elements that 'interest was aroused' in the PL (1964: 896–897). For Brush, PL 'attracted little attention' until the novel confirmations were obtained. (1996: 617) Mendeleev himself, in a letter published in 1879 to the Chemical News, said that the successful novel confirmations had served to 'direct attention' toward the periodic law – suggesting once again that many chemists had not given it much attention prior to those confirmations.

My position is that the periodic law was ignored by a large body of imperfect experts for the same sorts of reasons that were identified in the discussion of such experts in Chapter 2 – the amount of time and effort that would be required to fully understand the theory and its attendant evidence was considerable (a point that will be substantiated more fully below), and prior to the novel confirmations there was no particular reason to think that such time and effort would be rewarded. After all, for all any imperfect expert knew, the theory was weakly supported or would be repudiated on some respectable basis at some point in the near future. D. M. Knight offers essentially this diagnosis for Stage 1 indifference to PL: "There are after all a lot of cranks about, and the professional scientist is reluctant to go on wild goose chases; and for this reason Mendeleev's predictions aroused little interest when they were first made." (1992: 137)

What drew attention to PL were primarily its astonishing novel confirmations. But why would novel confirmations be such attention getters? It essentially suffices to answer this question to show that many scientists among the relevant body of theory evaluators were pluralists – and this is clearly the case. In the case of imperfect evaluators, I suggest that the analysis of predictivism provided in this chapter can be applied to show and explain the shift in opinion toward PL. Novel successes were doubly blessed: they were invulnerable to the charge of endorsement cooking (such as the tempered unvirtue-ascribing evaluator might make) and they reduced doubts about whatever epistemic basis had prompted Mendeleev to endorse PL in his early work (such as the tempered virtue-ascribing evaluator might have). Novel confirmations could, unlike accommodations, escape the pluralist's suspicion of adhocery and confirmed the background beliefs of the endorser.

Stephen Brush makes a point that could serve as a criticism of the claim I am making about the novel success of a theory. He writes

As in other cases, such as the confirmation of Einstein's general relativity theory by the 1919 eclipse expedition, we have to distinguish between the value of a successful novel prediction in *publicizing* a theory, thus forcing scientists to give it serious consideration, and its value as *evidence* for that theory in comparison to other evidence. (1996: 610)

Brush sees, as I do, a commonality between the case of the General Relativity theory and that of the periodic law with respect to their novel successes – but his point seems opposed to the point I am making here.[10] For I have construed the attention-getting role of the novel confirmations as due to their propensity to confirm the theory for a pluralist evaluator – while Brush implies that their attention-getting role may not be evidential but simply one of publicizing a theory – of getting scientists to consider it. It is important to remember that Brush does argue for the truth of predictivism in this episode – but he implies that predictivism would not hold for an evaluator who had not yet fully understood the theory or the evidence that supports it. Confirmation, he may be suggesting, may occur

---

[10] Scerri (2007: 147) agrees that prediction may indeed serve the publicity function Brush attributes to it.

in the mind of an evaluator only after that theory has been fully grasped – but prior to that point novel confirmations may serve the role of gaining publicity for a theory.

But I reject Brush's distinction between publicity and confirmation. For one might ask why novel confirmations in particular have the ability to secure the attention of the scientific community if not for reasons of their confirming power. Why did Mendeleev's many accommodations fail to secure attention for PL? The confirming power of successful prediction, moreover, is easily understood from the standpoint of a pluralist evaluator, a point I need not further press. Brush's position may derive from a commitment to epistemic individualism in his view of actual theory evaluation, a view I have been at pains to discount. Once we accept that much theory evaluation is pluralistic, it is a short step to accepting the reality of tempered predictivism.

Another form of pluralist evaluator discussed in the previous chapter is the reflective evaluator – the evaluator who considers what evidence actually supports her own expert beliefs. Though an expert within a particular specialty may know the epistemic basis for some or perhaps many of her expert beliefs, it is a virtual certainty that she does not know them all. But if an expert is of a reflective character, she may wonder what basis she has to hold them nonetheless – or she may feel compelled to offer justification for her scientific beliefs to science skeptics of one sort or another. I suggested in Chapter 2 that the most reasonable course for the reflective expert is to adopt the perspective of the pluralist. She can point to the consensus that was reached by scientists in her specialty who were fully familiar with the relevant evidence, just as a non-expert could. But she could also point to the novel successes that practitioners of her field have generated, for such novel successes testify (given assumptions given in detail in this chapter) to the truthlikeness of the background beliefs that led to them. Now it seems to have been the case that chemists during and after Stage 2 pointed to the success of Mendeleev's predictions for just this purpose. This seems to have been Brush's point in the passage quoted above to the effect that "For some chemists, the success of Mendeleev's predictions meant that their science had at last attained the respected status long enjoyed by astronomy and physics. . . . the cases of gallium, scandium, and germanium were compared to that of Neptune as demonstrations of the capacity of mature exact science." (Brush 1996: 616)

The reference to the successful prediction of the existence of Neptune by astronomers as a 'demonstration' raises the question – demonstration to whom? One obvious answer is to the scientific community at large – including non-chemists – one that was indicative of the veracity and overall adequacy of chemistry as a theoretical field of inquiry. These would include experts who worked on fields that were related to chemistry, the sort of pluralist evaluator deemed in Chapter 2 an 'interdisciplinary expert.' Thus the quotation suggests another form of expert pluralism in the case of Mendeleev. But another answer would be to those chemists who had shared some or all of Mendeleev's background beliefs – insofar as they were of a reflective ilk, and prone to wonder about their own doxastic commitments, i.e., their own background beliefs. Insofar as those background beliefs had led to the endorsement of a theory that enjoyed stunning novel successes, those background beliefs were confirmed for both reflective and interdisciplinary evaluators.

The one remaining form of expert pluralist discussed in Chapter 2 is the humble evaluator. The humble evaluator suffers (at least in her own mind) not from any actual lack of knowledge or expertise but from a worry that she has, in her own act of estimating some theory's probability, committed some form of cognitive error. This could consist of failing to apply some relevant background belief in the course of her deliberations about some theory's probability, or committing some kind of computational error in that estimation. This form of epistemic humility, I have argued, is certain to be more prevalent in epistemic contexts in which theory probabilities are sensitive to a large and complex body of background belief and data. Humble experts are prone to be pluralist evaluators for the obvious reason that they regard their own probability estimations as fallible – and thus confirmed by the judgments of other competent experts who agree with them, or disconfirmed by competent dissenters.

It is important to remind ourselves at this point how precisely a humble evaluator will tend toward predictivism. Like any pluralist, the humble evaluator will look upon novel confirmations of theories as having a special tendency to confirm theories for all the reasons discussed so far: they transcend skepticism about the acts of the unvirtuous endorser and about the background beliefs of the virtuous endorser (or about the logical acumen and background beliefs of either type of endorser). This will apply to any endorser who posts an endorsement level probability for a theory. Now in the case of the evaluation of any particular theory we could think of the humble evaluator as a skeptic who posts a sub-endorsement probability for a theory that has been endorsed by others, or as an endorser who posts an endorsement level probability. In the former case, novel confirmations will confirm the judgment(s) of the endorser(s). Now they will do the same in the latter case as well – it is important to emphasize, however, that one of the endorsers will be the humble evaluator herself – even if she is the lone endorser. For the humble evaluator is the sort of person who adopts a skeptical attitude toward the accuracy of her prior probability based on whatever accommodated evidence she has considered. Accommodated evidence suffers from an intrinsic inability to assuage this sort of self doubt – to appeal to accommodated evidence as though it supports a theory will typically require that she trust that her own deliberations are error-free. But novel successes confirm her judgment in a qualitatively different way – they serve as a test of those deliberations and, when successful, testify to the accuracy of those deliberations.

While I cannot identify any particular scientist as a clear example of the humble evaluator in the case of Mendeleev, my position is that the claim that such evaluators existed is a plausible one. One clear fact about the periodic law is that the theory itself and the evidence that Mendeleev mustered on its behalf were both very complex. The actual body of empirical evidence that drove Mendeleev to commit himself to PL was gigantic. It was noted above that the version of PL offered in the 1871 paper was extremely complex; Mendeleev himself noted in the 1879 letter referred to above that the relevant phemonena are "very complicated." The likely result of this is that many readers who made a serious attempt to follow the arguments for PL – even one who found the arguments compelling and who accepted Mendeleev's background assumptions – probably ought to have adopted a healthy skepticism about whether she (along with Mendeleev) had committed some form of cognitive error in judging that PL (or some of its components, at least) was highly probable on the basis of the proffered evidence. The effect of this would be to motivate the pluralism, and hence the predictivism, of the humble evaluator: novel confirmations confirm the claim that the assessment of accommodated evidence was error-free. This should hold whether the humble evaluator is or is not herself an endorser of PL. I thus regard the existence of humility-based predictivism among the evaluators of Mendeleev's PL as quite likely.

Before bringing to a close this rather extended application of the account of predictivism articulated in this chapter and the last, some retrospective clarifications are in order. These involve the basic concept of endorsement and the associated asymmetry between accommodation and prediction vis à vis the confirmation of background belief for virtuous predictivism. In the first part of this chapter (3.4.2) I emphasized that the impotence of accommodated evidence to confirm background belief is essentially a matter of definition – for L is fixed in such a way that endorsers of T whose prior is at or above L are those whose credibility is enhanced by confirmation of further empirical consequences of T from the standpoint of some pluralist evaluator. Endorsers whose priors fall below L are those whose credibility is not enhanced by novel successes of T (cf. 2.2) – by definition of L. However, when we turn to Mendeleev and the case of the individualist virtue-ascribing evaluator (3.8.1), it turned out that accommodated evidence sometimes did (3.8.1.1 and 3.8.1.3), and sometimes did not (3.8.1.2), confirm background belief. But this is curious – how could accommodation confirm background belief, if its impotence in this regard is essentially an analytic truth? To see the answer we must consider again the basic concept of endorsement. The definition of L is, first of all, tailored to the pluralist evaluator. From the standpoint of such an evaluator, accommodated evidence has a certain in principle incapacity to confirm endorser credibility. This is because, without detailed information about the content of the endorser's reasoning and background beliefs, the pluralist has no reason to regard an endorser's accommodated evidence, however much of it there is, as by itself confirming the truth of the endorser's background beliefs. This is true whether the pluralist assumes that the endorser is virtuous or unvirtuous (though it is particularly obvious in the latter case). Proposing a theory which is based on a large amount of evidence, in other words, does not by itself offer much information that one is a credible authority (viz., possessed of true background beliefs). Thus, for the pluralist evaluator, accommodated evidence is in general impotent to confirm background belief. This is true despite the perfectly real possibility that accommodated evidence may, in some way, confirm the accommodator's background belief.

Matters are thus different for the individualist evaluator, as she is acquainted with the content of the accommodator's background beliefs and with the logical relationships between them and the salient evidence. Thus of course the individualist can tell if accommodated evidence does confirm such background beliefs. Nonetheless, as we saw in (3.8.1) there are multiple cases in which the individualist will discern an asymmetry between the endorser's predicted and accommodated evidence – these are cases in which accommodated evidence serves, when combined with background beliefs, to make a theory sufficiently plausible such that only at that point do subsequent confirmations of the theory's empirical consequences serve to elevate the probability of those background beliefs. The effect of the accommodated evidence, in short, is to turn the predicted consequences of T into consequences of the endorser's background beliefs (to some significant degree), and thus serve to confirm such beliefs. Matters are complicated even more by the fact that the distinction between accommodated and predicted evidence, for the individualist evaluator, may not coincide precisely with the line as defined by the endorser's endorsement activity (see, e.g., 3.8.1.1). Clearly, there is room for more discussion of the relationship between pluralist and individualist virtuous predictivism – but I have done all I can do here to sketch the relationship.

3.8.4 Competing theories of the Mendeleev case

In Chapter 1 we surveyed Maher's method-based theory of predictivism – Maher furthermore argues in his (1988) that his theory can illuminate the predictivist effect in the case of PL. Another theory that has been offered for the same purpose is that of Scerri and Worrall (2001) (developed further in Worrall 2005). In this section I consider both of these accounts and argue that my account of predictivism provides a superior account of the scientific community's evaluation of PL. I also briefly take note of how the theories of Kahn Landsberg and Stockman (1994), Lange (2001), White (2003), and Sober and Hitchcock (2004) might be applied to the case of PL – and why I prefer my own account to all of these.

My first point of criticism is that most of these theories are committed to the concept of use-novelty (exceptions are Worrall and White, cf. below). I have argued in Chapter 2 that use-novelty is misguided because it is based on a mistaken attribution of epistemic significance to the act of theory construction. Endorsement, not construction, is the epistemically significant act. Furthermore, as noted, the conception of use-novelty provides no way to explain that the extent to which a predictor actually predicts E can vary, depending on how high the predictor's prior probability for T is. This misses the important fact that prediction is a matter of degree: the more strongly a theorist endorses some theory that entails E, the more strongly that agent predicts E. The strength of the prediction is reflected in the phenomenon of per se predictivism – the higher the probability assigned to T by a predictor, the more strongly T is supported by her background beliefs (assuming a virtuous endorser). But let us consider each theory in some detail.

3.8.4.1 Maher: reliable methods of theory construction

Maher (1988, 1990) defends the claim that predictivism is true because successful prediction, unlike accommodation, confirms the claim that the predictor is using a reliable predictive method – in his (1993) he notes that one sort of 'predictive method' is any method of constructing hypotheses that entail predictions. Besides its commitment to use-novelty, Maher's theory suffers from several deficiencies that I believe are collectively fatal. First of all, Maher's theory of predictivism, in emphasizing methods of discovery, commits the same error as that which infects use-novelty – it mistakenly attributes epistemic significance to the act of theory construction. That a theorist constructs a theory might testify to the epistemic basis on which that theory stands, but only if construction is accompanied by a concomitant act of endorsement, and this is a contingent concomitance.[11]

Furthermore, Maher's account of predictivism is heavily committed to what I believe is an entirely controversial claim: that the construction of new theories is typically the result of what should be called the application of some method of theory construction. Prima facie, such a claim seems to count against any critical role for creativity and serendipity in theory construction – and this seems at odds with the usual idea that these factors often play an important role in discovery.

A good discussion of the role of method in theory construction is found in Kantorovich (1993: 49–60). Discovery methods vary greatly, he points out, in their generality – ranging from highly general (like 'Consider all the evidence and construct the best possible theory') to highly specific ('Use NMR spectroscopy to determine the structure of the molecule'). As a general rule, Kantorovich claims, the usefulness of a method is inversely related to its generality (1993: 51). The more specific the rule, the more straightforward its application – the less specific the rule, the more needful are creativity and imagination. But in many cases of discovery, particularly those that involve considerable conceptual innovation, only the more general method will be available to the theorist – thus considerable creativity will be required on her part. Kantorovich claims that it seems that creativity 'cannot be learned or instructed by method' (1993: 62).[12] Likewise Darden's (1991) identifies multiple strategies for theory construction used by the various scientists who collectively constructed the theory of Mendelian genetics, but these too are highly general (e.g., 'begin with a vague idea and refine,' 'make a simplifying assumption,' etc.). It is far from clear, given the literature on scientific discovery, to what extent the formulation of a new theory can be attributed to anything that ought to be called a 'method' of constructing hypotheses. To whatever extent new

---

[11] Maher also neglects the per se/success distinction, though his theory seems ultimately compatible with it.

[12] This is not to say, however, that creativity cannot be inculcated in theorists in various ways, as Kantorovich emphasizes – such as by acquiring detailed knowledge of the relevant field, being ready to depart from traditional ways of thinking, and applying ideas generated in one field to another (1993: 184–185).

theories are produced by highly general methods, moreover, how could such methods be assessed for 'reliability'? How reliable 'Begin with a vague idea and refine' is qua method would seem to vary tremendously depending on the skill and knowledge (and sheer industry) of the scientist using it.

Maher explicitly cites the example of Mendeleev in support of his theory. Mendeleev's method, he claims, "was to look for patterns in the properties of the elements" (1988: 276). Before the success of his predictions, Maher claims, skepticism about PL was partly attributable to skepticism about this method – it was not regarded as a reliable method of theory construction. But the success of his predictions substantially increased the scientific community's belief that this method was reliable. But I find this position curious in the following way: the method of looking for patterns in the properties of the elements was not a method used only by Mendeleev, or by the other scientists who participated in the formulation of PL – it was also used by those chemists who rejected PL (like Berthelot, for example). For those chemists also looked for patterns in the properties of the elements – they must have done so, insofar as they judged there to be no such patterns to be found. Nor is it clear that they regarded the method as unreliable – insofar as the use of this method led them to what they considered to be a true claim ('There are no patterns among the properties of the elements.') If the use of this method led only a few people to a true theory, but even more to a false theory, on what basis can one claim that the method is reliable rather than unreliable?

By contrast, the theory of predictivism I propose here avoids entirely the red herring of construction and the obscurity of appeals to method reliability. How theories are constructed, the extent to which construction is method driven, and what constitutes method reliability are entirely beside the point. My theory makes use only of very orthodox notions available within confirmation theory – such as the well-established claim that confirmation is a three-way relation between theory, evidence, and background belief. Predictivism emerges as a more straightforward and easily intelligible phenomenon of theory evaluation on the theory developed here – hence my preference for it.

3.8.4.2 *Worrall: conditional and unconditional confirmation*[13]

I will base my remarks here on Worrall's (2005) paper which purports to clarify and defend his theory against various criticisms I offer in Barnes

---

[13] In Barnes (2005) I offered several criticisms of Worrall's theory of predictivism and of Scerri and Worrall's (2001) attempt to apply Worrall's theory to the case of PL. Scerri and Worrall separately

(2005). Worrall first of all emphasizes that he does not actually accord epistemic significance to the issue of whether evidence was 'used' in the construction of theory despite the fact that he often writes as though this is his position:

Although presented as a version of the 'heuristic approach', [my theory] is at root a *logical* theory of confirmation – the important logical relations being between (i) the evidence at issue e, (ii) the general theoretical framework involved T, and (iii) the specific theory T' developed within that framework (and which (at least in the straightforward case) entails e). In the paradigm case, the chief question will be whether some parameter having a fixed value in T' was set at that value by theoretical considerations, or as a 'natural consequence' of such general considerations, or whether instead the value was fixed on the basis of the evidence (and if so whether the evidence needed to fix the value of the parameter was, or included, e itself). (Worrall 2005: 819)

Worrall shows how his account can illuminate the diminished evidential impact of certain accommodations in terms of an example involving creationism (cf. 1.4.4): suppose T represents the 'core idea' of creationism (the world was created by God in a single event) and e refers to the fossil record as it has been observed to be. The creationist can accommodate e by modifying T into T': God created the world in a single event and created the fossil record, just as it is observed to be, to test our faith. Now of course T' does entail e. But e has only a limited power to confirm T': if there is independent evidence that genuinely confirms T, then e confirms T'. But absent such independent evidence for T, e is powerless to confirm T'. This illustrates why, for Worrall, much accommodated evidence often has only a diminished ability to confirm theory. A prediction in this example would be some evidence that followed from T together with 'natural' or independently plausible auxiliary hypotheses – a successful prediction would thus serve to

---

authored replies to my paper. I attributed to Scerri and Worrall an attempt to show that predictivism did hold in the case of Mendeleev's predictions – I took them to be claiming that these predictions did offer more support than a similar quantity of Mendeleev's accommodated data (though they seemed to me to claim that the difference in evidential impact between predicted and accommodated evidence was smaller than other authors had claimed). Scerri responds at length that this was a misunderstanding on my part, given their attempt to "undermine [predictivism's] role in the historical case in question." (Scerri 2005) He emphasizes the importance of the case of argon as an example of the importance of accommodation, a point I consider and rebut above. However, this interpretation is hard to square with the fact that Scerri and Worrall's (2001) contains a detailed description of Worrall's theory of predictivism, along with the claim that at least some of the historical facts surrounding PL should be understood in terms of this theory (p. 427, p. 450) Worrall (2005) clearly intends to defend the application of his theory of predictivism to the case of PL – I address his remarks below – while claiming that the evidential importance of the accommodation of argon can be accounted for in terms of his theory. I am unsure how to reconcile the various claims of Scerri (2005) and Worrall (2005) about the thesis of their co-authored paper.

confirm T, and thus offer unconditional support for T' (assuming e is also known true, and T' represents the clearly best way to accommodate e).

Now Worrall (2005) clarifies how he would apply this model to the case of Mendeleev. I will summarize Worrall's position this way: let 'pL' refer to the general theoretical framework of the periodic law, viz., the claim that there is a periodic relationship between atomic weight and various chemical and physical properties of elements. Let 'pL*' refer to the specific version of the periodic law which incorporates all its claims about the specific atomic weights of elements together with all the detailed chemical and physical properties Mendeleev ascribed to the various elements. Now let 'O' refer to all the facts about atomic weights and physical and chemical properties of the sixty-two elements that Mendeleev took himself to know when he presented his theory at the beginning of Stage 1. Now Worrall explains that Mendeleev, when he formulated pL*, clearly did not need to make use of all of O. Some proper part of the information in O was all that Mendeleev needed to fix pL* – the remaining part of O served in effect as predictions entailed by pL* that were subsequently confirmed when Mendeleev checked the implications of pL* to determine whether they coincided with the remaining information in O. Thus Worrall is prepared to argue that in fact, prior to the discovery of gallium, scandium and germanium, pL* had actually enjoyed much predictive success in the minds of those scientists who appreciated these facts (including Mendeleev himself, of course). This reconciles Worrall's account of predictivism with the claim defended in Scerri and Worrall (2001: 413f.) that there was in fact a fair amount of support in the scientific community for pL* prior to the famous predictive successes mentioned above.

Let us refer to the portion of O that Mendeleev actually based his original version of pL* on as O1, and the remaining portion of O as O2. Worrall's application of his theory of confirmation to the case of Mendeleev thus comes down to two claims: (1) O1 was powerless to confirm pL, and offered only conditional confirmation of pL*, and (2) there is an asymmetry between O1 and O2 with respect to their ability to confirm pL*, as O2 offered direct support of pL, and thus provided unconditional confirmation of pL*. However, both (1) and (2) strike me as straightforwardly false. (1)'s plausibility suffers from the curious consequence that O1 managed to induce Mendeleev to propose pL, and pL*, while nonetheless offering no actual support to pL, and only conditional support to pL*. But in this case it becomes mysterious that O1 had the ability to induce Mendeleev to propose pL – why would one propose a theory for which there was literally no evidence? How did O1 manage to generate pL and pL* in Mendeleev's mind without offering at least some evidence for both pL and pL*? (2)'s plausibility suffers from the fact that there is no basis for any asymmetry between O1 and O2 in terms of their evidential support for pL or pL*. Recall Worrall's emphasis that what ultimately matters on his account is not whether evidence was used or not, but the logical relationships between evidence, the central idea of a theory, and the specific version of the theory. But the logical relationships between pL/pL* and O1 are exactly the same as the relationships between pl/pl* and O2.

The problem with Worrall's attempt to apply his theory of confirmation to the case of Mendeleev, in my view, is his assumption that the accommodated evidence O1 stands to pL/pL* essentially as the fossil evidence stands to the core idea of creationism theory and the specific version of creationism designed to accommodate the fossil record discussed above. But these two cases are utterly distinct. The fossil evidence clearly fails to confirm the creationist core idea – whereas O1 clearly confirms the core idea of the periodic law. Thus his attempt to argue that predictivism did hold – at some level, at least – in the case of the periodic law ultimately founders.

A comparison between Worrall's theory of confirmation and my own may be in order here. At a somewhat superficial level, Worrall's theory resembles my own. I have argued at length that predictivism is true, in some cases, because prediction confirms the background beliefs of the endorser more strongly than accommodation. Worrall holds that prediction can offer confirmation of the 'core idea' of a theory in a way that accommodation (sometimes) cannot. Now a 'core idea' sounds at least somewhat like a 'background belief.' But notice, first of all, that the kinds of belief I count as background go far beyond the core idea of a theory. I count Km1 and Km3 as Mendeleev's background beliefs – and neither of these remotely resembles 'core ideas' of PL (e.g., they make use of concepts, like 'Cannizaro's methods' and 'truth-conducivity' that form no part of PL). Km2, which claims that the domain of the elements forms some kind of unity, sounds like the sort of thing one might count as a core idea of PL, but I argued that (thin) predictivism actually failed in the case of Km2 (in one sense at least, cf. 3.8.1.2). The non-predictivism surrounding my Km2 resembles the non-predictivism surrounding Worrall's pL that I argue for above: accommodated and predicted evidence do about as well in both cases to support the relevant core idea/background belief.

3.8.4.3 Three other accounts

The two attempts at systematically illuminating the phenomenon of predictivism in the case of PL have been those of Maher and Worrall – I have explained why I prefer my own account to these. However, other theories of predictivism were canvassed in Chapter 1 which might be brought to bear on this case – and we should consider how they might fare. Kahn, Landsberg and Stockman (1992) argue that predictivism is true because prediction provides special evidence that the predictor is 'talented.' They could certainly point to the obvious fact that Mendeleev was both a predictor and indisputably talented qua scientist as evidence of the applicability of their theory to this case. One primary problem as I see it with KLS's account is that its fundamental concept of 'talent' is obscure – it amounts to an unexplicated tendency on the part of scientists to propose good theories. I prefer my own account because it is more specific: scientists who are prone to endorse novelly successful theories do so not simply because they are 'talented' but because they are possessed of true background beliefs or are avoiding ad hoc theory adjustments (for virtue-ascribing and unvirtue-ascribing evaluators, respectively). Lange's (2001) theory asserts that predictivism is true because prediction is correlated with a theory being a non-arbitrary conjunction – theories based on just accommodated evidence are somewhat prone to be arbitrary conjunctions. The notion of an 'arbitrary conjunction' is never fully explained by Lange – the critical point is that confirmation of one conjunct of an arbitrary conjunction provides no evidence for the truth of any other conjunction. So he might argue that the truth of Mendeleev's predictions provided evidence that PL was not an arbitrary conjunction. But this does not strike me as particularly promising – insofar as we can think of PL as a conjunct (of claims about atomic weights and elements properties) it was a non-arbitrary conjunction by definition – for it was a lawlike claim that asserted a nomological relationship between various element properties. An evaluator who worried that PL might be an arbitrary conjunction would have simply failed to understand the content of PL. This is not to say that it was true by definition that the confirmation of one 'conjunct' of PL provided actual support for the other conjuncts – for of course evaluators who were sufficiently skeptical about PL for independent reasons might find the confirmation of one conjunct reasonably powerless to confirm the others. But this would be the case because of the reasons for skepticism, not a misguided worry that PL was itself an 'arbitrary conjunction,' whatever this term means exactly.

Another theory is presented in White (2003), who argues that prediction, unlike accommodation, provides special evidence that the theorist was "reliably aiming at the truth" (664) or "was reliably hooked up to the facts" (669). It is not particularly clear how these locutions are to be interpreted, though there are passages in White's article that suggest he may have something in mind like my own account. For example, he claims that successful prediction may increase our confidence in the theorist's assessment of the relevant evidence (672), the reliability of the theorist's measuring instruments, the experts on which the theorist relied, the theorist's own perceptual faculties, etc. (673). But White also emphasizes that his position is NOT that the epistemic advantage of prediction consists in any special evidence it provides for the theorist's background beliefs, but for the "truth-aim of the methodology" (ft. 21 678), a phrase I am unsure how to interpret (White emphasizes in this passage that his version of the no-miracles argument for realism differs from Boyd's, cf. ch. 4). All this could be interpreted as a form of unvirtuous or virtuous predictivism (or both). White's account could be applied to Mendeleev: his successful predictions, unlike his accommodated evidence, provided special evidence that he was 'reliably aiming at the truth.' Ultimately White's theory suffers, in my view, from the same problem that some of the other theories discussed here suffer from: it is couched in terms of fundamental concepts that are fairly obscure.

One last theory of predictivism we should recall was Sober and Hitchcock's (2004) account that argued that predictivism was true because prediction was correlated with the absence of data overfitting – because (1) the process of theory construction did not involve accommodation of the predicted data, thus preventing the theorist from adjusting theory to fit the data more closely than a true theory would (given the presence of noise in the data) and (2) because successful prediction is evidence that overfitting did not occur. Sober and Hitchcock could also point to the fact that their account fits well with the case of Mendeleev – for Mendeleev did feel free to ignore a certain amount of accepted data because it did not coincide with the overall pattern identified by his theory (this led him to predict that certain accepted atomic weights, e.g., would eventually be corrected). He thus was both a successful predictor and was prone to avoid overfitting, at least in some cases.

Both Lange's and Sober and Hitchcock's theories of predictivism are theories that view accommodators with a suspicion that predictors escape – their theories are thus in roughly the same ball park as the account of unvirtuous predictivism I sketch above (and with Lipton's 2004 Ch. 10 account). But all these accounts fall short insofar as they ignore the phenomenon of virtuous predictivism – in which accommodators and predictors are treated as equals in logical acumen. In general, they underestimate the varieties of predictivist evaluators that can constitute a scientific community.

3.9 CONCLUSION

I have, in this chapter and the last, proposed to reconstrue the basic idea of a novel confirmation in terms of endorsement, and argued that predictivism can take various forms. The bulk of my analysis has concerned the form I call virtuous predictivism, according to which predictivism has two roots, one of which consists of the predictive act itself, the other of the effect of predictive success. Virtuous predictivism can hold for both individualist and pluralist evaluators – producing both a thin and tempered virtuous predictivism.

In closing this chapter I would like to return to a topic discussed in Chapter 1 – Popper's proposed distinction between science and pseudo-science. As everyone knows, Popper identified a scientific theory as one that made specific, falsifiable predictions, while a pseudoscientific theory did not, either because it made no predictions but only offered ex post facto accommodations, or because it made predictions which, when falsified, could be revoked so as to accommodate recalcitrant data ad hoc. Part of the appeal of this solution to the demarcation problem was obviously that a falsifiable theory could, if false, be shown to be false by making the appropriate observation, unlike an unfalsifiable theory. Popper's anti-inductivism prevented him from claiming that a scientific theory's predictions, if confirmed, strongly supported the theory, but the intuition that this is true on the part of many probably helped explained the appeal of Popper's solution to the demarcation problem. But there is another palpable intuition about falsifiable theories whose predictions are confirmed that explains this appeal, and that is that the people who endorse such theories show signs of knowing, in general, what they are talking about. The theory of predictivism developed here lends credence to this intuition.

CHAPTER 4

Miracle arguments and the demise of strong predictivism

4.1 INTRODUCTION

Once again, let us begin with the fantasy of you suddenly becoming rich, and seeking financial advice. You choose an advisor with a long string of successful predictions about which investments will be successful over another who can merely accommodate the data with his theory. The successful predictions testify to the greater credibility of the predictor, you say. But how confident should you be that the predictor's theory is actually true? Could the predictor, say, have a false account of the causes of market fluctuations which for some reason happens to systematically generate true predictions? In this chapter we consider the role of predictive success in arguing for the truth of theory.

Thus far we have approached this problem in two ways: in Chapter 2 I developed a general case for the prevalence of pluralism in scientific theory assessment and argued that pluralism provides a straightforward rationale for predictivism. In Chapter 3 I proposed a taxonomy of predictivisms explicated partly in a Bayesian framework. But predictivism is a sufficiently complex topic that it can be and has been approached from still other perspectives in philosophical literature. Another of these perspectives is supplied by the longstanding realist/anti-realist debate. As is well known, there is a long tradition stretching back to William Whewell of defending realism by appealing to the epistemic significance of novelty (cf. 1.4.1), a tradition that remains very much alive today. This tradition claims that realism – which incorporates the claim that the theories of the so-called mature sciences are at least approximately true – can be defended by pointing not simply to the empirical successes of contemporary theories but specifically to their novel successes. The primary argument offered by proponents of this tradition is a particular version of the miracle argument for scientific realism, which claims that the attribution of truth to contemporary theories is the only way to avoid the consequence that

their considerable novel successes are simply miraculous. Our reason for
joining this debate is not primarily for the sake of taking a stand on the realist/
anti-realist debate itself, but because it presents us with a particular account of
why novelty carries special epistemic import – one which is both very different
from the account developed thus far in this book and, as I shall show,
irretrievably flawed. Exposing this flaw is the first task of (4.2) below. This
flaw moreover corresponds to a flaw in the resulting version of the miracle
argument for realism, and exposing this is the second task of (4.2). Section 4.2
concludes by noting that the account of predictivism developed in the
previous chapters escapes the problem which afflicts the flawed account,
and by noting the direction that proponents of the miracle argument must
take in order to avoid the flaws of the traditional version of that argument.

   But there is another reason why we cannot avoid the account of predictivism that (4.2) is concerned to repudiate – and that is that this account is
the primary argument that has been given for strong predictivism. Strong
predictivism, as characterized in (1.5), holds that predictivism is true because
prediction in and of itself carries epistemic significance – rather than because
prediction is merely correlated with some other epistemic feature of theories
which carries such significance. Strong predictivism has been a popular
position among many proponents of the version of the miracle argument
for realism discussed below. But the argument they give for strong predictivism is flawed for essentially the same reason that the version of the miracle
argument they give is flawed. These issues will be explained below – my
conclusion will be that strong predictivism is completely unjustified.

   In (4.3) I take up the task of developing the surviving version of the
miracle argument for realism. I argue that the prospects for this argument
are reasonably good if a particular controversial assumption is true, one
which involves the rate at which empirically successful theories are
endorsed in the history of science. Because I offer no defense of this
assumption, my defense of realism is a moderate one.[1] But there is an
important consequence of this discussion, which is that the anti-realist
cannot embrace virtuous predictivism as the realist can – despite what
appears to be the position of anti-realists like van Fraassen and others on
this point.[2] This result forces a modification of the nature of virtuous
predictivism: novel success cannot count as evidence of the empirical

---

[1] As will become clear, however, the realist's position is not untenable even if the controversial
   assumption is not known to be true.
[2] I am not suggesting that van Fraassen or other proponents of constructive empiricism have explicitly
   endorsed predictivism, but simply that arguments that he (and others) have advanced could render it
   possible for the anti-realist to do so. I repudiate such arguments below.

adequacy of an endorser's background beliefs unless it is also evidence for the truth of those background beliefs. Thus an anti-realist who denies that novel success is evidence for the truth of background beliefs cannot hold on to virtuous predictivism by claiming that novel success is evidence for the empirical adequacy of such beliefs. The virtuous predictivist must be a realist.

In (4.4) I consider the critique of the realist/anti-realist debate presented in Magnus and Callender's (2004) article – they argue that many participants in the debate are guilty of committing the base rate fallacy. I argue that their defense of 'retail realism' suffers from its failure to appreciate points developed in this chapter.

Finally, in (4.5) I point out that the philosopher who adopts realism is properly regarded as another example of an epistemic pluralist who clearly is committed to the epistemic significance of the judgments of scientists. Ultimately, the dispute between the realist and the anti-realist comes down to a dispute about these judgments. The position of the realist philosopher who adopts the version of the miracle argument defended in this chapter is essentially that of the non-expert (2.3.2) or pluralistic expert (2.4). This clarifies the nature of the realist/anti-realist debate.

Before proceeding a few preliminary clarifications are in order. Realism, as noted, includes the view that the theories of the so-called mature sciences are at least approximately true. Realists also typically hold that scientists have the ability to systematically identify and accept theories that are approximately true. Anti-realism is a skeptical position that repudiates the possibility of theoretical knowledge and holds that the empirical success of theories is no reason to regard them as approximately true. I will understand the anti-realist to hold that ordinary claims about observable objects are often knowable. I want to modify a distinction due to Psillos (1996: 34) and call 'horizontal inference' the inference that moves from premises about observed entities to conclusions about unobserved but observable entities. 'Vertical inferences' move from premises about observed entities to conclusions about unobservable entities and their properties (such as those of electrons, ions, quarks, etc.) The anti-realist at issue here accepts as truth-conducive the ordinary practices of horizontal inference but rejects the truth-conducivity of vertical inference (or at least rejects the claim that one must posit vertical truth conducivity to explain anything one observes in the history or practice of science).[3] The realist accepts as truth-conducive both the ordinary

---

[3] It appears that van Fraassen himself does not entirely endorse this claim, as he now claims that he does not accept the truth conducivity of horizontal inference to the best explanation (cf. Ladyman et al., 1998). I agree with the judgment of Richmond (1999) that this represents a serious liability for van

practices of horizontal inference and the practice of vertical inference as practiced by the scientific community.

The realist/anti-realist debate has undergone considerable evolution over the years. One historically important form of anti-realism was embraced in the early twentieth century by the movement known as logical positivism which argued for an anti-realist interpretation of scientific theory based on the verifiability theory of meaning. The logical positivists claimed that a scientific theory was logically equivalent to a conjunction of statements couched entirely in a vocabulary of terms referring to observable entities. Something (very roughly) in the same ballpark as this kind of semantic anti-realism was championed by Quine (1951, 1960). Early defenders of realism such as Maxwell (1961) rejected semantic anti-realism and argued for epistemic realism, which claims that there is evidence that justifies beliefs in the literal truthlikeness of theoretical statements. The realist/anti-realist debate, however, underwent a significant change with the publication of van Fraassen's (1980), which presented a version of epistemic anti-realism he called 'constructive empiricism.' Constructive empiricism rejected semantic-antirealism, i.e., it allowed for a literal interpretation of theoretical statements, but embraced epistemic anti-realism. In the decades that followed constructive empiricism emerged as the most viable form of anti-realism, and it is the general type of anti-realism at stake in this chapter. From this standpoint, the realist/anti-realist debate is primarily a debate about whether the evidence scientists muster on behalf of theory merits the kind of high probability judgments associated with belief.

4.2 NEITHER TRUTH NOR EMPIRICAL ADEQUACY EXPLAIN NOVEL SUCCESS

The miracle argument for scientific realism has taken various forms over the years. One form simply claims that we should regard contemporary scientific theories as approximately true because that is the only way to explain their astounding empirical successes, where 'empirical successes' are simply understood as their true empirical consequences. This form accords no explicit significance to novelty. I will not be particularly concerned with this form in this chapter. Instead, I will focus on a popular version of the miracle argument for realism that does accord special status

Fraassen's current position, and do not identify the common version of anti-realism I discuss here with the totality of van Fraassen's latest views, though of course the version I discuss is generally attributed to van Fraassen (1980).

to novelty – and focus also on another argument I call the miracle argument for strong predictivism. These arguments are tightly intertwined. They are also, I will argue, subtly but fatally flawed. This is because both arguments involve a fallacious application of Ockham's razor. My intention is not to repudiate realism or, of course, predictivism, but to identify an argument for realism that does not commit this error and to note that the account of predictivism expounded in the previous chapters does not run afoul of this error. The primary conclusion I will argue for in this section is that the popular attempt to explain why a theory is novelly successful cannot be explained by attributing truth – or empirical adequacy – to the theory. To explain novel success requires a fundamentally different type of explanation. Before proceeding to these issues, a brief discussion of the principle of Ockham's razor is needed.

4.2.1 The anti-superfluity principle

The version of Ockham's razor that concerns us I deem the 'anti-superfluity principle,' or ASP (cf. Barnes 2000). ASP maintains that putative explanations should not be endorsed when they are explanatorily superfluous vis à vis the truths they purport to explain. For example, one could deploy the ASP in arguing that there is no reason to endorse a claim C_1 which posits an overdetermining cause of E when another claim C_2 which posits a sufficient cause of E is known true (even though both 'C_2' and 'C_1&C_2' explain E equally well and thus might – naively – be thought to be on equal footing). Let E = My cherry tree has fallen over, C_1 = John cut it down, and C_2 = James cut it down. If John confesses but says nothing about James, there is no reason to assume that James added his labor to John's, even though the claim that it was the result of both their labor explains the explanandum as well as the claim that John acted alone. The point is trivial where C_1 and C_2 are mutually exclusive, since knowledge of C_2 in this case entails the falsehood of C_1. But where C_1 and C_2 are compatible (but non-overlapping and uncorrelated) putative explanations of E, and where C_1 is by itself sufficient for the truth of E, the truth of E provides no evidence for C_2 given independent knowledge of C_1's truth. So C_2 should not be endorsed on the basis of the truth of E (though of course it might be endorsed on some other basis) – this is what the ASP claims.

I want to draw attention to a terribly obvious fact about the ASP. To deploy the ASP against some claim C_1 which purports to explain some data D, it must be the case that there is some claim C_2 which is known true and which is an adequate explanation of the *same* data D. For example, to deploy the ASP against the claim that James cut down the cherry tree, assuming that the only data that supports this claim is the fallen tree, one must know to be true some other adequate explanation for the fact that the tree has fallen over. One could not deploy the ASP against the claim about James by claiming to know some claim which does not even apparently explain this explanandum (e.g., I may know that the tree in question was planted by my grandfather, but since this fact does not even apparently explain why my tree was cut down, such knowledge does not render the claim that James cut it down superfluous for explaining the explanandum). Let us refer to the requirement that there must be a known explanation that explains the *same* fact as the allegedly superfluous explanation as the 'sameness condition' on the ASP.

4.2.2 *The miracle arguments for realism and strong predictivism*

The miracle argument for scientific realism is usually attributed to Putnam (1975: 73), and claims that realism is the only position that does not make the empirical successes of scientific theories a miracle. Many proponents of this type of argument, however, have been quick to point out that these empirical successes are of course no miracle insofar as these theories were built to fit the true observation statements they entail. What realism is needed to explain – on one version of the argument – are the novel successes of theories – where 'novel' is clearly explicated as 'use-novel.' Theories with substantial records of use-novel success thus ought to be regarded as probably true – for these successful predictions stand in need of explanation, and the attribution of truth to these theories is both an adequate explanation and the best explanation of this success. But when an explanation of some phenomenon is both adequate and the best of the competition, it ought to be regarded as probably true – QED.

In a thorough review of the literature on the miracle argument for realism up to 1988, Alan Musgrave points out that "careful realists, beginning with William Whewell, distinguished two kinds of predictive success, predicting known effects and predicting novel effects ... According to Whewell, Duhem and Popper, then, what is really surprising or miraculous about science, what really needs explaining, is novel predictive success, rather than predictive success *simpliciter*." (1988: 232–234) (Although Duhem is generally considered an arch-anti-realist, Musgrave shows that Duhem conceded that the miracle argument for realism – couched in terms of novel predictions – "had some force.") Musgrave acknowledges that other important proponents of realism like Putnam and Boyd do not emphasize novel success "as one should" (1988: 234–235). The basic idea that one may defend realism by pointing out that realism is the only legitimate explanation for the *novel* successes of theories is the central thesis of Leplin (1997) and is also defended by Psillos (1999). Thus Musgrave's 'final version' of the miracle argument for realism: "It is best construed as an inference to the best explanation of facts about science. The facts which need explaining are best construed as facts about the *novel* predictive success of *particular* scientific theories [italics in original]. The realist explanations of such facts are best construed as invoking (conjecturally) the truth of those theories (or their near truth if we can develop such a notion)." (1988: 249)

Before proceeding to show what is wrong with this argument, a word of clarification is in order. As noted, the realists cited by Musgrave spell out their argument for realism in terms of use-novelty. I have defended a new conception of predictive novelty in this book which I argue is superior to use-novelty. However, for the moment the dispute over how to conceptualize 'novelty' is not the primary issue before us. For the time being I will adopt the usual assumption that novelty is use-novelty – as we will see, this is important to understanding the problem with the orthodox miracle argument sketched above – and the related orthodox argument for strong predictivism. Consequently, the claim that N (a consequence of T) is a novel confirmation of T amounts to the claim that N has been shown true but T was not built to fit N, viz., N was not used in the construction of T. However, ultimately the version of the miracle argument (and predictivism, of course) I want to defend will be spelled out in terms of my notion of endorsement-novelty.

A more rigorous formulation of the relevant version of the miracle argument for realism is as follows:

Let (F) be the fact that a theory T entails a conjunction of true observation statements E:
a. (F) stands in need of explanation.
b. There are just two possible adequate explanations: (1) T was built to fit E and (2) T is true (for if neither (1) nor (2) are true then the truth of F would be a 'miracle' and thus would have no explanation).
c. Now if (1) is known to be true, then we have an adequate explanation of (F) – and there is no need to endorse (2). (ASP)
d. Where (1) is known to be false, then we should endorse (2), since it is the only adequate explanation of (F) available.
e. Thus T should be regarded as probably true if T was not built to fit E.

In what follows we will refer to this as the miracle argument for realism. But it is important to note that the argument just given not only counts as an argument for realism – it is also an argument for strong predictivism.[4] For from (a–d) it follows that (f) T is more strongly con-firmed when T is not built to fit E than when it is built to fit E – we will refer to this argument as the miracle argument for strong predictivism. This argument, like the miracle argument for realism, argues that only the imputation of truth to a novelly successful theory blocks the consequence that novel success is miraculous. All proponents of the miracle argument for realism referred to above implicitly endorse this argument for strong predictivism – they are essentially the same argument.

Both arguments critically apply the ASP at premise (c). Just as one should not endorse the conjunction that John and James cut down the tree when the claim that John cut down the tree is both known true and sufficient to explain the fallen tree, so one should not endorse the con-junction that the theory was both built to fit E and also true when the claim that T was built to fit E is both known true and sufficient to explain the fact that T entails true E. Superfluous explanations are unsupported by the data they purport to explain – this is, again, the point of the ASP.

Notice that a proponent of the miracle argument for strong predictivism need not claim that evidence that a theory is built to fit never confirms the theory – for as we have seen there are multiple cases in which it does. The point is that if all we know about some theory T is that it entails some true observation statements that it was built to fit, T less strongly confirmed by those pieces of evidence than if it had not been built to fit those statements.

My thesis is that both miracle arguments incorporate a subtle but fatal flaw. Both arguments assume that one may explain the fact that a theory is novelly successful by imputing truth to the theory. Despite the plausibility of this claim, I will argue that it is false. But the critique I will give applies with equal force to the claim that one may explain a theory's novel successes by attributing mere empirical adequacy to a theory (as an anti-realist might hope to do). My objective in this section is primarily to undermine two apparently plausible arguments – the miracle arguments for realism and strong predictivism summarized above.

My strategy will be as follows. I will begin by considering 'built to fit' explanations of theory success. My first objective is to demonstrate that one cannot explain why a theory has a true empirical consequence by asserting that the theory was built to fit that consequence (thus premise (b) of the

---

[4] The predictivism entailed by this argument is strong because it clearly entails the truth of biograph-icalism: knowledge of theorist's intentions in theory building is essential to a complete understanding of the evidence that supports the theory.

miracle arguments is false). The attribution of truth (or empirical adequacy), however, does explain why a theory has a true empirical consequence. This means that the built to fit explanation and the truth explanation explain different things. Consequently the attempt to deploy the ASP against the truth explanation, when the built to fit explanation is known true, violates the sameness condition on the ASP. Hence the miracle arguments for realism and predictivism as stated above should be rejected.

However, I will go on to propose new versions of these miracle arguments that respect the sameness condition on the ASP. While one cannot explain why a theory has a true entailment by applying the built to fit explanation, one can explain why a particular theory was endorsed by asserting that it was built to fit some known conjunction of observation statements. My notion of 'endorsement' is the broad one developed in Chapter 2. Built to fit explanations, I will argue, are explanations not of truth entailment facts (as the miracle arguments sketched above presume) but of endorsement facts about theories. But insofar as a built to fit explanation is supposed to show that some other explanation A is super-fluous by the lights of the ASP, A must also be an explanation of an endorsement fact – this is what the sameness condition on the ASP requires. Now A will be the sort of explanation to which we turn when the built to fit explanation is not available – thus A will be an explanation of novel success. Thus explanations of novel success must also be explanations of endorsement facts – and not truth entailment facts. Thus neither truth nor empirical adequacy explains novel success – for one cannot (in general) explain why a theory was endorsed by attributing either semantic property to that theory. Only the imputation to a theorist of a set of background beliefs that is true or at least empirically adequate can serve as a candidate explanation of novel success – for these sorts of explanations have the resources to explain why theories with true novel observational entailments are endorsed when such theories were not built to fit these true observational statements.[5] Proponents of realism (who emphasize the epistemic significance of novelty) and predictivism can offer such explanations. But the resulting predictivism turns out not to be strong predictivism but weak predictivism – for the special probative weight of prediction is now simply

---

[5] My argument that explanations of novel success are explanations of endorsement facts should not be taken to imply that we have at this point relapsed to my definition of 'novel success' in terms of endorsement-novelty. Novelty at this point in our argument is use-novelty – but insofar as built to fit explanations explain endorsement facts rather than entailment facts, explanations of novel success (empirical successes of theories not built to fit such empirical truths) must also explain endorsement facts if the sameness condition on the ASP is to be respected. More anon.

a matter of a correlation between successful prediction and true back-ground beliefs. Thus we are led back to the theory of virtuous predictivism developed in the previous chapter. Let us begin by considering 'built to fit' explanations more carefully.

4.2.3 What do 'built to fit' explanations really explain?

According to both miracle arguments, the fact that a theory T entails a particular (possibly conjunctive) observation statement E can be explained by noting that T was built to fit E. I will describe the fact that T entails some statement as an 'entailment fact' about T. One can designate entail-ment facts about theories in two ways. I use the term 'E-entailment fact about T' to refer to the fact that T entails some determinate statement(s) E. I use the term 'truth-entailment fact about T' to refer to the fact that T entails some true statement(s). Truth entailment facts can be constituted by entailments that are specified or unspecified. So, an E-entailment fact about Einstein's general theory of relativity is that it entailed the descrip-tion of the Mercury perihelion. A specified truth-entailment fact about this theory is that it entailed the true statement describing the Mercury perihelion – an unspecified truth-entailment fact is that it entailed some true statement(s) or other. One might attempt to explain an E-entailment fact about T by asserting that T was built to fit E, and one might attempt to explain a (specified or unspecified) truth-entailment fact about T by asserting that T was built to fit the true observation statements T entails.

A typical example of such reasoning is Leplin (1997: 31): "... how a theory manages to predict ... phenomena does not require its truth for explanation, if the phenomena are assumed in constructing the theory." The same point is made in Zahar (1989: 14–15). Likewise Musgrave (1988: 231): Regarding the argument of Clavius that the predictive success of Ptolemaic astronomy would be 'incredible' if that theory were not true, "This simply can be denied. After all, Babylonian astronomers detected periodicities in astronomical phenomena and devised algebraic rules for predicting them. It is hardly incredible or miraculous that a rule expressly devised to capture some periodic phenomenon should successfully predict future instances of that periodic phenomenon." The point is that one can explain why the 'rule' (read: theory) entails a true observation statement by citing the fact that the rule was built to fit the relevant phenomenon. Schlesinger attributes to the 'great majority of philosophers' the view that the highest test of a theory is to "ask it to indicate in advance things which the future alone will reveal." After all, "anyone of us can, with the benefit of hindsight, easily construct ad hoc hypotheses to fit results we already know." (1987: 32) The clear implication is that 'construction with benefit of hindsight' suffices to explain the fact that a theory has true implications. Redhead (1986: 117) claims that when h is built to fit e then the "explanation of e is guaranteed independent of whether h is true or false." Giere (1984: 159–161) argues that a theory that was built to fit some body of evidence is not supported by that evidence because the theory stood no chance of being refuted by that evidence. The implication is that a theory built to fit E is thereby guaranteed to entail E – thus built to fit explanations explain entailment facts (the same point is made in Glymour (1980: 115)).[6]

Now it strikes me as rather obvious that neither E-entailment facts nor truth-entailment facts can be explained by claiming that a theory was built to fit the relevant observation statements. For the claim that some theory was built to fit some observation statement is a claim about a contingent event – an event in which some theorist formulates some theory on some particular basis. But entailment facts are either logical truths or contain logical truths as components. But logical truths cannot be explained by contingent events! Suppose (f1) T entails e – then f1 is a logical truth and contingent events (like those leading to the construction of a theory) cannot explain it. Suppose (f2) T entails true e. Now f2 is not a logical truth, for it entails the truth of e. But (f2) is equivalent to the conjunction of (a) T logically entails e and (b) e is true. But neither conjunct is explained by the fact that T was built to fit e – (a) is a logical truth and (b) cannot be explained by some description of how some theory was built (it is explainable only by some theory like T). But if the built to fit explanation explains neither (a) nor (b) then it does not explain the conjunction of (a) and (b) either.[7] Finally, suppose (f3) T entails an unspecified true observation statement O. (f3) is essentially an existential generalization that claims that there exists an observation statement O such that T entails O and O is true. But the built to fit explanation does not explain this existential claim either. The reason is simply that the identity and the truth-value of all T's observable consequences – whether specified or unspecified – is entirely unaffected by which process was used to construct T.

---

[6] Howson (1990: 229) points out in response to Giere that whether H was built to fit e cannot affect the 'chance' that e refutes H. Howson does not go so far as to argue that 'built to fit' explanations cannot explain entailment facts – the point I am about to press – but he clearly makes a point in the same ballpark.

[7] I do not claim that logical truths cannot in principle be given some kind of explanation. For example, Kitcher (1989: 422–428) claims that logical and mathematical truths can be explained in terms of Kitcher's unification theory of explanation. I claim here only that such truths cannot be explained by postulating contingent events, a point consistent with Kitcher's account.

It seems intuitively clear, however, that built to fit explanations are not entirely without any explanatory force – the above analysis raises the question what built to fit explanations do explain, if not entailment facts about theories. My position is that built to fit explanations serve to explain not why a particular theory has true observable entailments, but why a theory with such entailments was endorsed by a theorist. My account of 'endorsement' is, as noted above, the broad account developed in Chapter 2 and deployed in Chapter 3: a theorist endorses a theory insofar as the scientist performs a selection procedure that results in some degree of endorsement (however minimal) of that theory by that theorist. This selection procedure could consist of the process of declaring that a theory has some degree of plausibility – one that could range from 'being worthy of further consideration' to 'being highly probable.' Consider a theory T which some theorist builds to fit some determinate observation statement E. The fact that the theorist knew E to be true, and desired to have an E-entailing theory, led her to construct a theory that would have that entailment – and then led her to endorse T (in the broad sense). The theory building process that incorporates the accommodation of E culminated in the endorsement of T. Thus built to fit explanations serve to explain endorsement facts about theories, and not entailment facts. Alternately, and somewhat more accurately, since there are multiple theories besides T that would entail E, the fact that the theorist built her theory to fit E explains why she endorsed some E-entailing theory (additional explanatory information will be needed to explain why T was the theory she endorsed rather than some other E-entailing theory, unless this selection of T from the class of E-entailers was essentially random).[8]

Heretofore we have considered cases in which T logically entails some observation statement – so of course we are imagining that T is a conjunction of what are usually called theoretical hypotheses together with what are called auxiliary hypotheses (theoretical hypotheses by themselves

---

[8] Mayo (1996: 266) makes a distinction which somewhat resembles my distinction between explanations of entailment facts and explanations of endorsement facts. With respect to some hypothesis H that has been built to fit true E, we may ask why H accords well with the evidence – answer (1) is that ANY hypothesis that was built to fit E in such a way would end up being a hypothesis that entails E, while answer (2) might be that H is error free. Mayo is not arguing that built to fit explanations do not explain entailment facts, as I am. In fact, the wording in (1) suggests that she may think that how a theory was built can explain why it has the entailments it has. She is concerned that the conflation of the two interpretations of the question may lead us to erroneously conclude that all hypotheses that are use-constructed are illegitimate. Insofar as answer (1) is generally available to explain why H fits E without positing the truth of H, this might lead us to erroneously reject (2). I concede of course that hypotheses can be well confirmed on the basis of evidence they were built to fit.

entail no observation statements). But there is another type of case – suppose that T is simply a conjunction of theoretical hypotheses (e.g., suppose T is the hard core of a Lakatosian research program). Now T by itself carries no observational consequences. Nonetheless, we might claim that T entails – in a loose sense – some observation statement O insofar as auxiliary hypotheses A have been endorsed by some agent(s) X such that T & A logically entails O. Let us describe this state of affairs by asserting that T 'loosely entails' O for X. Now the fact that T loosely entails O for X is not a logical truth – since X could have endorsed auxiliaries other than A. Could one thus explain loose entailments in terms of built to fit explanations? It would seem so, since the fact that T loosely entails O for X could take as its explanation the fact that auxiliaries were 'built' for the very purpose of allowing T to loosely entail O. Alternately, the fact that T loosely entails O for X could be explained by noting that T was constructed by X because it was clear that the conjunction of T and A would entail O (where A was antecedently accepted on some independent basis) – or perhaps both T and A were built because it was clear that their conjunction entails O. However, it is important to consider what kind of a fact a loose entailment is. While it appears to be a kind of entailment fact, it is more accurately thought of as a kind of endorsement fact. For the fact that T loosely entails O for X is equivalent to the fact that X put forward (either tentatively or decisively) some hypothesis or hypotheses such that the resulting conjunction logically entails O. (The fact that T loosely entails O cannot be equated with the fact that some auxiliaries exist such that their conjunction with T entails O, since for any T and any O, there are auxiliary hypotheses A such that T & A logically entails O (e.g., for a trivial case, let A be a logical falsehood).) So while the built to fit explanation can serve to explain a loose entailment, it turns out that in doing so it is explaining an endorsement fact. Once again, it turns out that built to fit explanations should be construed as explanations of endorsement facts about theories.

But if my argument is right, a considerable mystery remains. Why has the claim that one may explain entailment facts with 'built to fit' explanations seemed so plausible to so many for so long? There are, I believe, a variety of reasons for this popular misconception, and it is to this question that I now turn.

One issue that strikes me as relevant to our current concern is what kind of thing one takes a theory to be. Suppose, for example, we were to assume that theories are abstract entities that subsist eternally in some Platonic realm. On this approach, there is something misguided about describing a human theorist as 'building' a theory to fit a certain body of evidence – insofar as this locution implies that the theorist brings the theory into existence by her own cognitive activity. The theorist in this case is better described not as 'building' a theory but as simply being the first to formulate a theory that has always existed. So 'building a theory to fit E' simply involves 'formulating a theory because that theory entailed E.' So the theorist's intention to accommodate E and subsequent creative activity serves to explain the fact that the theory was conceived by that theorist – not the timeless logical truth that that theory entails E.

But now suppose that we adopt a conceptualist ontology of theories, and regard theories not as abstract entities but as mental representations which have no existence apart from their concrete instantiation in human minds. On this approach the locution of 'theory building' appears to make better sense, for human theorists do literally bring the theory into existence by their own creative cognitive acts. Theories are thus similar in this regard to human artifacts like cars and computers. Now it is quite plausible to claim that the properties artifacts possess typically are possessed as a matter of contingent truth, and that the explanation for their possession of those properties consists in facts about how the artifact was built. That my Toyota Corolla has the property of being the right size to fit inside a typical lane of traffic is properly explained in terms of the car-building process that includes the intention of designers to make sure the car will fit therein. But of course the claim that my Toyota fits inside a typical lane of traffic is not a logical truth. Logical truths about my Toyota (like "This Toyota is white or non-white") cannot be explained in terms of the car-building process. So – to some degree – the intuition that E-entailment facts about theories can be explained in terms of the theory-building process arises on the basis of a faulty analogy between theories and their logical properties and artifacts and their non-logical properties.

4.2.4 Back to the miracle arguments – and the critical appeal to Ockham's razor

As noted, the miracle arguments for realism and strong predictivism make use of the ASP. Both arguments hold that there is no reason to regard a theory as true on the basis of the fact that that theory has true observational entailments when that theory was built to fit those entailments. Insofar as the 'built to fit' explanation is known to be true, the postulation of the theory's truth is supposedly rendered explanatorily superfluous. But in order to run these arguments in good conscience, it seems to me, one must be in the grip of the illusion that built to fit explanations are explanations of truth entailment facts. For the sameness condition requires that the competing explanations be putative explanations of the same facts. Now the imputation of truth does indeed explain why a theory has true entailments (either unspecified or specified as 'true E') – so to deploy the ASP against the imputation of truth on the basis of the theory's being built to fit the known data requires that one take built to fit explanations to be explanations of truth entailment facts. But at this point we know that built to fit explanations are not explanations of truth entailment facts. So both miracle arguments are fatally flawed insofar as they fail to respect the sameness condition on the ASP.

Can Ockham's razor be deployed to argue for realism or predictivism in a way that respects the sameness condition? Insofar as 'built to fit' explanations are explanations of endorsement facts, the question is how endorsement facts are to be explained when built to fit explanations are not available, i.e., under circumstances of an endorsed theory's being novelly successful. But this is to inquire how a scientist managed to endorse a theory that entailed true observation statements without building the theory to fit those statements. Assuming the endorsement was not a lucky accident and admits of some explanation, the relevant candidate explanation is that that scientist had a set of background beliefs K that were approximately true or at least approximately empirically adequate. The attribution of such K to such a theorist helps explain how that theorist endorsed a theory that turned out to have true novel observational consequences. When the scientist did, on the other hand, build the theory to fit such statements, there is no need to attribute to the scientist a set of true or empirically adequate background beliefs to explain the endorsement fact, and thus no particular reason to regard the theory as true or empirically adequate beyond the true observational statements it is otherwise known to fit.

To restate the argument somewhat more simply: one cannot explain why a theory has a particular true empirical consequence E by appealing to the fact (supposing it to be one) that the theory was constructed so that it yields E. Instead, what the fact that the theory was constructed to yield true e explains is the 'endorsement' of that theory (or of that particular version of the theory). But then the miracle arguments need to be recast – because the truth of a theory DOES explain why it entails true E, and hence – since the two explanations explain different things – they cannot be rival explanations as the miracle arguments (as presented above) suppose. Realism and predictivism should be defended by miracle arguments that claim that only the imputation of an appropriate set of background beliefs serve to show that novel success – the endorsement of novelly successful theories – is not miraculous.[9]

So, while one may – in the context of the miracle arguments – explain novel success by attributing an appropriate set of background beliefs to the theorist who proposed the novelly successful theory, one may not – in that context – cogently explain novel success by imputing truth or empirical adequacy to such a theory (on pain of violating the sameness condition). For all that has been said, this latter conclusion may still seem strange to some readers. The antidote for this strangeness is to think carefully about what one attributes to a theory in attributing use-novel success to it. A theory is use-novelly successful when it entails true observation statements that it was not built to fit. But in this case to attribute use-novel success to a theory is by that very attribution to make a claim about how that theory came to be endorsed – so to explain novel success one must explain, in some sense, how such a theory actually came to be endorsed in that way. Attributions of truth (or empirical adequacy) to the theory itself are explanatorily useless in this respect. While they may explain success, they cannot explain novel success.[10]

Larry Laudan noted long ago that the expression 'the success of science' is critically ambiguous (1984). 'The success of science' could refer to the success of particular theories in generating true observable consequences

---

[9] I am indebted to a referee at the *Australasian Journal of Philosophy* for this formulation of my argument.

[10] Commenting on an earlier and less clearly written version of this argument, a referee at the *Australasian Journal of Philosophy* (who I deem AJP Referee #2, as he/she is not the same as the AJP referee referred to in ft. 9) argues that I have misidentified the relevant explanandum of the miracle argument for realism. Contrary to what AJP Referee #2 believes is my position, "The realist does not employ the success argument to explain why P is entailed by T when P is deductively entailed by T. What the realist success argument explains is something very different, which presupposes the entailment fact that T entails P. What the realist seeks to explain is that T is empirically successful, where this means that the predictive consequences (e.g., P) it entails are true." I agree completely that no proponent of the miracle argument holds that the relevant explananda are the logical facts that T has certain entailments – such logical facts I deem above 'E-entailment facts.' It seems to me that AJP Referee #2 is essentially in agreement with the position I attribute to most proponents of the miracle argument in holding that the relevant explananda are truth-entailment facts. For to explain why T is empirically successful is just to explain its truth entailment facts. I see little difference between inquiring why such truth entailment facts hold of T and inquiring why certain observation statements are true while 'presupposing' that T entails them. If AJP Referee #2 means that what is 'presupposed' literally forms no part of the explanandum, then he or she is suggesting that the success of science is constituted merely by the truth of certain observation statements, and this strikes me as implausible. Ultimately, of course, I want to argue that, for those who hold to the epistemic significance of novelty, the relevant explananda ought to be endorsement facts about theories.

(irrespective of how those theories were chosen) but could also refer to the success of scientists in endorsing theories that subsequently proved empirically successful. Laudan's distinction thus shows that there have always been two versions of the miracle argument for realism to be considered – according to one version, realism is the only position that does not make the empirical successes of particular theories miraculous (since realism imputes approximate truth to such theories) – let us call this 'the miraculous theory argument for realism.' According to another version, realism is the only position that does not make the fact that scientists have been successful in endorsing theories that subsequently proved successful miraculous (since realism imputes empirical reliability to the scientific method of theory appraisal) – let us call this 'the miraculous endorsement argument for realism.' Although Laudan's distinction is rarely acknowledged explicitly in the vast literature on realism (for an exception, see Stanford 2000), realists have actually developed both of these versions. Of the two, the vast majority of literature is devoted to the miraculous theory argument.[11] But there has been an important proponent of a certain version of the miraculous endorsement argument in Richard Boyd (1981, 1991) – who argues that only realism can explain the instrumental reliability of method, i.e., the success of scientific method in producing theories that turn out empirically successful. Boyd argues that theory appraisal consists of choosing theories that cohere with background belief – insofar as the realist regards such background belief as approximately true, she can explain the instrumental reliability of method – theories that cohere with true beliefs are themselves likely to be true (and thus empirically successful). It should be noted, however, that Boyd makes no mention of novelty at all – he defends the version of the miracle argument that treats all empirical successes on a par. Nonetheless, Boyd does endorse a version of the miraculous endorsement argument. Although plenty of skepticism abounds about realism in the philosophical community, overall it appears that these two versions of the miracle argument are regarded by most discussants as more or less equally plausible. My position is that only one of these versions is viable – the miraculous endorsement argument – for

[11] This fact is attested to both by recent work on the miracle argument and by earlier literature reviews of the field. Fine's 1986 'state of the art' essay surveys the debate over the 'explanationist defense of realism' at length but clearly understands this as just what I call the miraculous theory argument (despite a cursory reference to Boyd's work (162–163)). Musgrave's 1988 literature review is quite similar in focus. As of the late 1990s little had apparently changed: Leplin's (1997) is almost entirely devoted to the development of the miraculous theory argument; extended anti-realist replies to Leplin are found in Kukla (1998 Ch. 2). Stanford (2000) clearly marks the distinction between the two forms of the miracle argument but goes on to focus just on the miraculous theory argument.

those who want to appeal to the epistemic significance of novelty. In my view, the prospects of realism are tied to a considerable degree to the eventual success or failure of this argument. In the next section I take up the question of whether this argument can succeed.

What then of predictivism? The miracle argument for strong predictivism argues that truth explains novel success – thus novel success argues directly for truth. This is what makes this argument an argument for STRONG predictivism, because strong predictivism claims that novel success is in and of itself epistemically relevant. And if novel success is a direct form of evidence for theory truth, then one must know whether evidence is novel in order to accurately assess the evidence for that theory – thus biographicalism holds, and the paradox of predictivism is fully with us.

But novel success is not direct evidence of theory truth – it is rather evidence for the credibility of the endorser. More specifically, it is evidence for the truth or empirical adequacy of the endorser's background beliefs. Novel success thus counts as indirect evidence for theory truth – it argues for endorser credibility, which in turn can be evidence for theory truth. But to re-conceive the argument for predictivism in this way is to re-conceive it as an argument for weak predictivism – for now novel success argues for truth just because novel success is a symptom of something else which is, in certain contexts at least, epistemically relevant: this is the credibility of the endorser. The credibility of the endorser is epistemically relevant in contexts in which evaluators are pluralists who lack perfect prior information about endorser credibility. The predictivism we are left with is the tempered predictivism of Chapters 2 and 3. This too leaves us with biographicalism, but of the benign sort associated with pluralism. Where evaluators are individualists, we are left with the thin predictivism of Chapter 3. This version of predictivism corresponds to an argument for predictivism (the miraculous endorsement argument for predictivism) that does not violate the sameness condition on the ASP.[12]

Thus, neither truth nor empirical adequacy explains a theory's novel success. However, it is the case, I believe, that both truth and empirical

[12] Although I criticize the versions of predictivism developed by Maher (1988, 1990, 1993), Kahn, Landsberg and Stockman (1992), Lange (2001) and Sober and Hitchcock (2004) in the previous chapter, it should be noted that these versions escape the critique I apply against the miracle argument for predictivism in this chapter. This is because these accounts do not explain novel success by simply attributing truth to theory, but by positing a theory construction process that is in some sense reliable – thus they purport to explain the 'construction' of novelly successful theories. Likewise Mayo (1996: 267) argues that the prohibition on use-construction is based on a desire to avoid unreliable methods of theory construction. I of course would re-construe their arguments so as to explain not the construction of such theories but their endorsement.

adequacy are plausible explanations of empirical success. But how can this be? It is tempting to think of a theory's novel successes as simply a subset of its empirical successes – so any explanation of a theory's general empirical success should suffice as an explanation of its novel successes. But this is specious reasoning. The critical point is that to describe a theory as 'empirically successful' is simply to assert a set of truth entailment facts about a theory – viz., that its observable consequences are all true. To describe a theory as novelly successful is to assert an endorsement fact about a theory – i.e., it was endorsed on some evidential basis that did not appeal to all its true observable consequences. Explanations of success thus differ in kind from explanations of novel success.[13]

In this chapter thus far I have assumed the orthodox notion of novelty qua use-novelty – which claims that a novel confirmation of a theory is a true empirical consequence the theory was not built to fit. But it is time to recall that the conception of novelty that I ultimately want to defend is endorsement novelty. On the latter notion, a non-novel confirmation of a theory is not one that the theory was built to fit, but rather one that the theory was endorsed (at least in part) because it fit. But just as one cannot explain why a theory has a true entailment by describing how the theory was constructed, neither can one explain why a theory has a true entailment by describing the basis on which the theory was endorsed. Above I argued (on the assumption that novelty was use-novelty) that built to fit explanations are, contrary to widespread belief, explanations of endorsement facts. More obviously still, 'endorsed because it fit' facts also serve as explanations of endorsement facts. The various arguments and analysis in the chapter thus far can be easily translated into the terms of endorsement-novelty rather than use-novelty.

4.3 THE MIRACULOUS ENDORSEMENT ARGUMENT FOR REALISM

In this section I take up the version of the miracle argument that I defend above: the miraculous endorsement argument that claims that realism is the only position that does not make the fact that contemporary theories have substantial records of novel success miraculous. In the next section I discuss the history of this argument up to Boyd's development of

---

[13] White (2003) argues in a very similar way that truth cannot explain novel success, though his analysis does not emphasize the role of Ockham's razor as mine does. I originally presented this argument in Barnes (2002a).

his version, and propose an improved version of this argument that emphasizes the epistemic significance of novelty. I go on to develop two objections against the improved version and defend the argument against them – and conclude that this version is defensible on a certain controversial assumption.

But before proceeding it is worth pausing to explain why the development of this version of the miracle argument for realism is important for present purposes. The general explanation is that the details of the realist/anti-realist debate cannot be avoided by anyone seeking a thorough grasp of predictivism. As we have just seen, the fallacious nature of the miraculous theory argument for realism was intertwined with a flawed argument for strong predictivism. In a similar way, my defense of the miraculous endorsement argument will carry consequences for the theory of virtuous predictivism defended in this book. Specifically, it will turn out that only the realist can embrace virtuous predictivism – and to a degree unvirtuous predictivism as well. My primary objective in exploring the miraculous endorsement argument is to substantiate this important point about predictivism rather than to defend realism per se.

4.3.1 The current state of the miraculous endorsement argument for realism

Many years ago van Fraassen provided a Darwinian explanation of the empirical successes of contemporary scientific theory (1980: 39–40). Such success can be explained by noting that those theories which are accepted today would have been discarded had they failed to survive the rigorous tests to which they were subjected. Thus it is no surprise that currently accepted theories are theories which have shown themselves empirically successful.

Laudan noted that the Darwinian explanation was at best only partially successful (1990: 55) (see also Musgrave 1985: 210). For while the Darwinian explanation does explain why theories currently endorsed (to the point of acceptance) by scientists have shown themselves to be empirically successful (thus far), it does not explain why those particular theories (irrespective of why they were endorsed) proved successful.

Laudan's distinction is an important one to bear in mind as we consider the current status of the miracle argument for realism. For, once again, it is rarely noted that there are two versions of the miracle argument corresponding to Laudan's two explananda. These are the two versions of the miracle argument described in the previous section. The Darwinian explanation explains how thus far empirically successful theories have come to be strongly endorsed on the assumption that they were proposed (viz., they were never falsified) and thus counters the miraculous endorsement argument rather than the miraculous theory argument. Boyd countered that the analogy between Darwinian theory and the history of science failed (1985: 23–28). As Darwin himself conceded, his theory would have been refuted if it had been the case that genetic variation was not random, i.e., if newly produced variations in species systematically favored organisms which were well suited to the environment. But the history of science reveals (according to Boyd) that newly emerging theories turn out to be empirically successful at a rate which exceeds what one would expect if the proposal of new theories were indeed as random as the anti-realist claims it is. Just as the phenomenon of directed genetic variation would have refuted Darwin's non-teleological theory, so the high rate of successful theory emergence refutes the claim that the success of science can be accounted for entirely in terms of the forces of 'natural selection' applied by the testing process to theories (cf. also Gutting 1985: 123). Thus Boyd argues, as I would put the point, that the Darwinian explanation fails to show that the anti-realist can render the endorsement of empirically adequate theories non-miraculous.

I do not know whether the rate at which newly emerging theories turn out to be empirically successful exceeds what one would expect if the process were as random as the anti-realist claims. But if it does not, the Darwinian explanation clearly suffices to defeat the miraculous endorsement argument for realism. The only question that remains is whether the realist can win the miraculous endorsement argument if the rate of successful theory emergence does exceed what would be expected if theory proposal were random in the anti-realist's sense. In what follows, we will assume that this rate is too high to be accounted for by chance – bearing in mind that we have by no means proven this or even argued it to be the case. I will refer to this assumption as the 'high frequency assumption.' How strong is the case for realism under the high frequency assumption?

Boyd explains that scientific method, broadly construed as including theory appraisal, experimental design, procedures of data collection, problem solution, etc., is highly theory-dependent (for an extended illustration of this claim see Boyd (1981)). The only plausible explanation of the instrumental reliability of our methodological principles, Boyd argues, is that the background theories K upon which scientific method is based are approximately true.[14] Thus the realist, unlike the anti-realist, can account

---

[14] We might stop at this moment to revisit the question which theories are in K, and thus serve to 'ground' scientific method. These theories include currently accepted scientific theories – these

for the fact that successful theories have been endorsed without the postulation of miracle: empirically successful theories were endorsed because the background theories by whose lights they were endorsed were true – and thus the process of theory evaluation was conducive to empirical adequacy.

Boyd's defense of realism raises an obvious question: how is the endorsement of the true theories K to be explained? Boyd says only that K's emergence was a contingent event (1991: 211). But now it seems to me that Boyd's sketch of the history of science is ultimately inadequate. Boyd's writings do not suffice to show that the realist can provide a picture of the history of science which leads to the acceptance of empirically adequate theories that is clearly miracle free. Nor is it at all clear that the anti-realist could not easily supply a competing account of the history of science that is just as miracle free as that provided by Boyd – if not more so. In what follows I develop these objections and then go on to show how the realist can respond to them.

4.3.2 Two objections to Boyd's miraculous endorsement argument for realism

The challenge facing the realist who would deploy the kind of miracle argument deployed by Boyd is to argue that her explanation of the instrumental reliability of contemporary scientific method (as constituted by the high frequency assumption) is not ultimately question begging. Her predicament resembles that of one who would explain why the Earth does not fall by claiming that the Earth rests on the back of a huge elephant – a proponent of this explanation immediately faces the challenge of explaining why the elephant itself does not fall. Of course one may always postulate another supporting elephant – but the addendum that 'it's elephants all the way down' fails to satisfy. Likewise the Boydian realist may explain the instrumental reliability of method at a particular point in the history of science by postulating true background theories at that point, but immediately faces the challenge of explaining how these background

ground the methodological claim that newly emerging theories should be judged plausible in part on the basis of whether they cohere with accepted theories. Furthermore, various types of scientific instruments used in research are built on the presumption that certain theories of the instruments are true – to assume that the instrument is reliable is to assume that these theories are approximately true. But K includes as well general metaphysical claims about the general nature of the physical world – these could include the claims that the true theories tend to meet various extra-empirical criteria. The claim that such extra-empirical criteria as deployed by scientists tend to change over time and reflect the changing fortunes of various theories is well defended by McAllister (1996). In all these senses, accepted scientific theories constituting K serve to ground methodological principles.

theories were endorsed. The addendum that 'it's true theories all the way back' in the history of science is equally implausible. So the Boydian realist faces the considerable challenge of providing a plausible starting point for the sequential generation of approximately true theories. Such a starting point must involve the generation of a 'first good theory' in some non-miraculous way.

The expression 'first good theory' should not be taken to imply that the realist is committed to there being a unique starting point for all contemporary true theories. There is no necessity in there being a unique first good theory from which all contemporary true theory descends (any more than it is necessary that there is a single organism from which all current life forms evolved – life may have arisen independently at multiple locations). But if the Boydian approach is to win the miracle argument, all contemporary theories judged probably approximately true by the realist must have ancestries which trace back to 'first good theories' (however many of these there may have been).

Boyd's (1981) makes a similar point: the approximately true theories of contemporary science must have ancestries that trace back to a 'take-off point' – a first point in the history of science at which a critical mass of approximately true theories had been accepted and could serve as a reliable constraint on subsequent theory choice. Let us call the various 'first true theories' that constitute a take-off point 'take-off theories.' According to Boyd, take-off theories emerge because they are favored by methodological principles grounded on background theories that were not approximately true. This entails, he argues, that it is not clear that the take-off theories should be considered as then constituting knowledge. Boyd's judgment about this is that it should not ultimately be the business of scientific epistemology to provide a neat classification of theories-at-moments as knowledge or non-knowledge (1981: 631).

But there is a more serious problem with take-off theories than whether they should count as knowledge, and that is that Boyd has not provided a miracle-free account of their emergence. For if a set of entirely false background theories grounds a set of methodological principles which somehow serve to recommend an approximately true theory, what could this be but a supremely lucky accident? What Boyd says (as in the above quotation, for example) is that the emergence of take-off theories was a 'contingent event' – so there was no a priori necessity that it did occur. But of course to admit this is not to show that the emergence of take-off theories was not miraculous – and this is what the realist must show. This is my first objection to Boyd's account of the history of science: he

provides no account of the emergence of take-off theories that is clearly
miracle-free. Without such an account, Boyd cannot claim to win the
miraculous endorsement argument for realism: he has simply buried the
miracle into the emergence of take-off theories. We take up this challenge
(deemed hereafter 'the problem of take-off theories') below.

The second objection I want to consider was proposed by van Fraassen
(1980: 80), Fine (1991: 83) and by Ladyman et al. (1997: 308).[15] Boyd argues
that the only plausible explanation of the instrumental reliability of contemporary method requires the postulation that contemporary background theories K are approximately true. But would it not suffice to
explain the instrumental reliability of method to ascribe mere empirical
adequacy to K? If K were not necessarily true, but empirically adequate,
then presumably the requirement that new theories be sanctioned by
methodological principles grounded on K would render it plausible that
new theories were themselves empirically adequate.[16] But this would
apparently suffice to explain the instrumental reliability of contemporary
method in a way consistent with anti-realism. Of course, the anti-realist
would face the same challenge faced by the realist: how is the endorsement
of empirically adequate background theories to be explained? But now the
position of the anti-realist seems to be perfectly symmetrical with that of
the realist – the anti-realist can explain the empirical adequacy of K by
claiming that K was judged plausible by a theoretical tradition in which the
constitutive theories were themselves approximately empirically adequate.
Now the anti-realist faces the problem of take-off theories, but it is hardly
obvious that the anti-realist will find this problem any more challenging
than the realist will. Boyd's claim that take-off theories simply emerged

---

[15] Ladyman et al. (1997) is a response to Psillos (1996), which essentially endorses Boyd's defense of
realism (which Psillos also attributes to Lipton (1991) and (1992–3)).

[16] It might be objected that from the assumption that K is itself empirically adequate it does not follow
that some newly proposed theory T which coheres with K is likely to be empirically adequate. But in
fact the empirical adequacy of K does entail this on a plausible conception of empirical adequacy.
This conception holds that to say of some theory T that it is empirically adequate is not merely to say
that those observation statements that follow with deductive certainty from T are true, but also to say
that those observation statements that follow with mere probability from T are probably true. But if
K is empirically adequate, and T is probable on K, then T's observational consequences are also
probable on K – thus theories that cohere with K are probably empirically adequate. (One should
not object to this claim by citing the well known point that the conjunction of two empirically
adequate theories is not itself necessarily empirically adequate. What is at stake at this point is not
whether an empirically adequate K will, when conjoined to a theory T that has been independently
found empirically adequate, yield an empirically adequate conjunction. What is at stake is whether a
theory T (whose empirical adequacy at this point is undetermined) which coheres with an empirically adequate K will itself likely prove empirically adequate. The answer to this question is yes,
I argue, on a plausible conception of empirical adequacy.)

'contingently' could just as well be deployed by the anti-realist to argue that take-off theories, which were empirically adequate but not necessarily true, emerged contingently. Given that the claim of empirical adequacy is weaker than the claim of truth for any theory, the anti-realist picture would render the emergence of take-off theories apparently less miraculous than the competing realist picture. Let us call this line of reasoning 'the anti-realist challenge.' In what follows I attempt to show how the realist can handle both the problem of take-off theories and the anti-realist challenge.

4.3.2.1 *The problem of take-off theories*

The problem of take-off theories is to describe how 'first good theories' could emerge without it being a miracle – a lucky accident – that such emerging theories are good. But what is meant by a theory's 'emergence'? The point of a theory's emergence, as I propose to use this term, should not be equated with the point at which it is regarded as unanimously endorsed. But neither should it be equated with the point in time at which some nascent version of the theory first occurred in some creative scientist's mind. A theory emerges, for present purposes, at the point that it is assessed by at least some competent scientists as sufficiently plausible to be worth further investigation – it thus receives a certain preliminary form of endorsement from the scientific community. So, for a theory to emerge, it must of course be formulated, it must be presented to at least some faction of the scientific community, and it must be evaluated by at least some competent scientists as possessing at least some preliminary plausibility on the basis of some evidence or argument(s) – and thus qualify as endorsed to that extent.

The task facing the realist is to argue that a first true theory's emergence need not be regarded as a lucky accident from the realist point of view. But now the realist seems to face a serious problem. We have all been told of Kekule's discovery of the benzene ring, which supposedly occurred while when he saw a flame in a fire that had a ring shape. It was, one might say, just a lucky accident that he was gazing into the fire at that moment and thus just an accident that the true theory about the structure of benzene emerged. But of course it need not be the realist's position that luck plays no role whatsoever in the emergence of theories, including take-off theories. What the realist must claim is that the considerations that led to the theory's being regarded as plausible by competent scientists were in fact truth conducive. If those considerations were not truth conducive (as the anti-realist will surely claim), then the fact (if it is a fact) that a true take-off theory came to be regarded as plausible was just a miracle. If those considerations were truth conducive, then it was no accident at all that a true take-off theory was deemed plausible. The problem is that Boyd's general answer as to what constitutes the truth conducivity of theory appraisal – the truth of background theories – is not available in this context. In fact, Boyd's insistence that all theory appraisal is theory-laden apparently guarantees that the emergence of take-off theories will have to be miraculous – in that it can only be a miracle if true theories are sanctioned by methodological principles based on entirely false theories.

What is the way out of this dilemma? My proposal is that the realist must argue that take-off theories were assessed as plausible on some basis that was not itself theory laden – but was nonetheless truth conducive. How is the realist to argue for this? It will do no good for her to argue that take-off theories are assessed as plausible simply because they fit a body of data – for it is always the case that indefinitely many theories can be proposed which fit any body of data. If data fitting is the only criterion, then all data-fitting theories must be deemed equally probable, and thus (assuming there are many such theories) no one theory can be assessed as plausible. Some extra-empirical criteria will be required to single out particular theories as plausible – but what are these criteria, and why would one regard them as truth conducive? Equally importantly, can any such criteria escape the charge of being theory-laden? My proposal is that take-off theories are sometimes judged plausible or promising, at least in part, because they provide an explanation of a salient body of data that meets some extra-empirical criterion (or criteria) that does not depend for its epistemic credibility on the acceptance of any scientific theory. Thus the preliminary confirmation of take-off theories requires some version of the assumption that theories, which meet extra-empirical criteria of this sort, are more likely to be true than competing theories which do not. The realist thus incurs the obligation to argue for the truth-conducivity of such principles in a way that does not appeal to the presumed truth of background theory – and it is my position that this obligation can be met, at least in some cases. It is not my position that there is just one such type of extra-empirical criterion, and there is every reason to suppose there will be many. But for ease of exposition I will focus below on the most commonly cited example of such a criterion, and that is the criterion of simplicity. The notion of 'simplicity' is one on which philosophers of science have spilt much ink. One currently popular view of what it means to claim that some theory is simple in an epistemically honorable sense is that that theory coheres with existing background theory (see, e.g., Sober (1988 Ch. 2). (In a similar vein McAllister (1996) argues on the basis of extensive historical data that various aesthetic criteria (other than simplicity) are accepted as truth conducive by scientists because certain already accepted theories display those aesthetic properties.) On such a view, for a theory to be simple is for it to simply meet the one extra-empirical criterion that Boyd identifies: coherence with background theory. While I agree that this is one important sense of simplicity, I reject the claim that all simplicity judgments involve implicit references to background theory. One clear reason for this (as I shall explain below) is that simplicity (in various senses of the word) clearly plays a role in inference at the strictly observable level – on the part of apparently rational agents who may not even be aware of the contents of any scientific theories. Candidate take-off theories can be assessed in terms of such 'atheoretical' simplicity criteria – and the realist can attempt to extrapolate the truth-conducivity of such criteria from horizontal to vertical inference.

To illustrate the emergence of a take-off theory I would point to the emergence of the early form of Mendelian genetics. A natural suggestion about when that theory emerged would be the publication in 1865 of Mendel's paper reporting his experiments on pea plants. In this paper he reported the property of dominance in pairs of related, phenotypic unit characters and reported the 3:1 ratios of dominant-to-recessive characters for seven traits of pea plants in the offspring of hybrid crosses. However, it has been widely claimed that this paper was utterly ignored until 1900 when Mendelian genetics was re-discovered and presented to the scientific public by three biologists. Darden points out that between 1865 and 1900 there were actually many citations of Mendel's paper in scientific literature. She writes, however, that "it still seems correct to say . . . that no one, before 1900, had published results showing that they had read Mendel's 1865 paper and appreciated both the significance of his numerical ratios and his explanation in terms of different types of germ cells." (1991: 44) Thus we would do better to pick 1900 as the year in which early Mendelian genetics emerged: it had at that point been clearly presented to the scientific public and had clearly elicited the judgement from some portion of that public that it was a plausible theory that deserved further scrutiny.

Needless to say, at this point Mendelian genetics was hardly a well-developed theory – as Darden's (1991) shows in detail, the various components of the theory evolved gradually over a thirty-year period. Mendelian genetics was, moreover, hardly uncontroversial in 1900. But we are not concerned with the point in time at which this theory was uncontroversially accepted – we are concerned with the point of its emergence as a new theory (one which many realists would now surely regard as 'approximately true' for many species). More specifically, we are concerned with the question how this theory came to be judged as plausible at this time. My position is that for the most part this theory was NOT assessed as plausible in 1900 because the theory was judged plausible by the light of methodological principles which were themselves theory-laden. Rather, the theory was assessed as plausible because it provided an explanation of a salient body of data that was strikingly simple in a sense that can be appreciated without appeal to accepted theory. Thus the early form of Mendelian genetics has the requisite features of a take-off theory – or so I shall argue in what follows.

The most important proponent of Mendelian genetics at this time was William Bateson – often called the founder of Mendelian genetics. Darden argues that there were various reasons why Bateson regarded the "new conceptions of Mendelism" as promising (1991: 46). The first of these was simply that the Mendelian conceptions allowed an explanation of a large variety of observed phenomena. These included Mendel's original observations of pea plants, but also the observations of de Vries and Correns of the 3:1 ratio in multiple species. It also included the observation that newly arisen variations were sometimes maintained in populations rather than being swamped out by the blending of phenotypic characteristics (insofar as it was a consequence of Mendelian genetics that some offspring of hybrids that differed phenotypically from the parent generation (the homozygous recessives) would 'breed true' thereafter). It is my position that Bateson's assessment of the plausibility of Mendelian genetics is rightly attributed – primarily – simply to the ability of the theory to account for this salient body of data. Of course more complex theories could have been devised that were consistent with this data – such as theories which posited different mechanisms for each species, theories which posited larger numbers of 'factors' than Mendel did in accounting for the 3:1 ratios, etc. What was striking about Mendelian genetics was its simplicity – in this example, such simplicity amounted to its unifying power and the paucity of its theoretical components. Thus the judgment that this theory was promising could be accounted for without requiring that any particular scientific theory of the day be accepted. Also, prior to learning about Mendelian genetics, Bateson was convinced that the path to a correct theory of heredity was through the techniques of artificial breeding and numerical analysis of data – that Mendelian genetics was based on experiments utilizing these techniques was yet one more reason for its appeal to Bateson – though not one apparently based on Bateson's acceptance of any scientific theory of his day. Neither the experiments that were designed, nor the observations that these biologists made, nor the inference to the Mendelian explanation itself were critically dependent on the acceptance of anything that deserves to be called a scientific theory. On the assumption (to which most realists should now be congenial) that Mendelian genetics was in some sense approximately true, Mendelian genetics qualifies as a take-off theory.[17]

Having said this, let me hasten to add that an additional part of the motivation for Mendelian genetics was theory based. For an additional reason the theory was judged promising by Bateson was its promise to resolve certain problems pertaining to evolutionary theory – to which Bateson was committed. One problem facing evolutionary theory was to account for how new species could arise – but the ability of Mendelian genetics to account for sharp differences between the traits of parent and offspring was promising in this regard. So it is not my claim that Bateson's assessment of this theory was entirely 'theory-free.' Rather, it is my claim that the theory-free component of the motivation for Mendelian genetics was by itself sufficient to explain the theory's being regarded, by Bateson and others, as plausible and worth further investigation – and thus sufficient to establish Mendelian genetics circa 1900 as a take-off theory.

Examples of what realists can regard as take-off theories can be easily multiplied. Consider Kuhn's conception of a paradigm, Lakatos's analogous conception of a research tradition, Laudan's conception of a research program, etc. Such theoretical traditions of course always have starting points – nascent versions of theories which are subsequently refined and improved over the course of the tradition. Consider whatever theoretical traditions have persevered to the present day, and which have generated theories that have proven astonishingly successful empirically – the ones the realist wants to argue should be regarded as approximately true. The earliest versions of such theories are obviously not judged promising because they cohere with methodological principles which are grounded on the presumed truth of the theoretical tradition they replace. How then are they determined as promising? My position is that such early theories

---

[17] Darden (1991: 31) actually lists eleven 'steps' that presumably preceded Mendel's original conceptualization of the basic components of his theory – three of these steps could be regarded as background assumptions that facilitated the discovery of the theory and the preliminary judgment that the theory was plausible. These include the construal of characters as units, not as parts of the 'essence' of the species, assuming that the account for one pair of characters could be extended to account for two pairs when 9:3:3:1 ratios were found, and making a general assumption that biological phenomena can be analyzed mathematically.' None of these assumptions were unreasonable in an exploratory context, but none of them, as far as I can see, were grounded on the presumed truth of any other scientific theory.

are often judged as initially plausible primarily because they account for a salient body of data in a strikingly simple way. Consider, e.g., the earliest arguments for the earliest version of Copernican astronomy, which consisted primarily of the argument that heliocentrism provided a simpler account of the observed motions of heavenly bodies than Ptolemaic astronomy. In this case, claims about the relative simplicity of Copernican astronomy over Ptolemaic astronomy have been the subject of some controversy. It is often claimed, e.g., that Copernican astronomy required no fewer 'circles' than Ptolemaic astronomy. But Copernicus' scheme is systematically simpler in the sense that its various component concepts are 'deductively interlocked' (see Edwards (1967: 221–222)). Here again the realist can find, I believe, the requisite atheoretic notion of simplicity. Some sixteenth-century scientists judged Copernican theory promising not because of its coherence with existing theory but in spite of its radical incoherence with most sixteenth-century theory (of mechanics, for example). Of course it may be true that some surviving theoretical traditions did not emerge in this way – but because their earliest constituent theories were favored by methodological principles which were grounded on background theories K that were uncontroversially accepted (though not on the theories of any competing theoretical tradition). But in this case the realist must simply argue that the ancestry of K traces back to the emergence of take-off theories – theories which the realist argues are approximately true and which emerged because they were sanctioned by methodological principles that were theory free – or at least had some theory-free component which was itself sufficient to account for the emergence of the theory. What matters is that the realist can show – at least in most cases – that strikingly successful theories have histories that are miracle free in the sense that there are no critical moments at which some ancestor of the successful theory emerged for reasons that were not at all truth conducive (according to the realist). I have argued that it is at least plausible that the realist can do this.

If the realist is to point to any particular theory as an example of a take-off theory, she must maintain (1) that the theory was approximately true, (2) that the reasons for its emergence were sufficiently theory-free, and (3) that the reasons for its being assessed as plausible were in fact truth conducive reasons. Anti-realists will naturally refuse to grant (1), but it is important to remember that the realist in this context is not defending the possibility of theoretical knowledge per se to the anti-realist, but is arguing that realism can provide a picture of the history of science that is miracle free – from the standpoint of realism itself. I have argued for the truth of

(2) above in the case of Mendelian genetics. What then of (3), which the anti-realist is likewise destined to reject? My position is that the ability of take-off theories like early Mendelian genetics to account for a salient body of data in a strikingly simple way is sufficient to explain how this theory came to be regarded as 'promising' by its early proponents – but for the realist, to be 'promising' means to possess some non-negligible probability of being approximately true. Once a theory has been accorded some such probability, it would seem to simply be a matter of providing a sufficient body of new evidence to drive the theory's probability to some level sufficient for justification.[18] To accord any take-off theory such a probability requires noting that the theory does explain a significant body of data and that it clearly meets some extra-empirical criterion for theory assessment – which one can plausibly argue to be truth conducive.

So, as usual, the realist owes us an argument for the truth conducivity of simplicity (or whatever extra-empirical criterion the realist requires). One common realist argument begins by reminding the anti-realist of their shared commitment to the truth-conducivity of common practices of horizontal inference. At the level of observables, realist and anti-realist agree that knowledge is often possible – including knowledge of unobserved but observable entities which is based on common practices of horizontal inference. But at the horizontal level, simplicity judgments clearly play a critical role in shaping judgment. To cite a rather worn out example, if I hear the pitter patter of little feet inside the wainscoting of some interior wall, and if I notice droppings characteristic of mice on the floor, and if food items left out of the fridge overnight appear nibbled, I would properly infer that a mouse is living in my house. Of course there are other theories which could be proposed which are equally consistent with this data – I could be the victim of an elaborate hoax played by someone who has a bet with someone else that he can convince me that there is a mouse in the wainscoting. But competing explanations like these suffer from an apparently dishonorable complexity. But if we agree that simplicity assessments are truth conducive at the level of horizontal inference, why not extrapolate this truth conducivity to the level of vertical inference – and thereby call oneself a realist? This is perhaps the oldest argument for realism in the book: inference to the best (in this case, the simplest) explanation is truth conducive at the horizontal level – so there is

---

[18] Hence van Fraassen's point that the anti-realist cannot assign non-0 point valued prior probabilities to theories and his proposal that the anti-realist should assign interval valued priors to theories whose lower bound is 0 (1989: 194).

every reason to regard it as truth conducive at the vertical level too. I am simply pointing out that this argument could be deployed on behalf of the approximate truth of take-off theories.[19] Anti-realists, needless to say, have multiple arguments against this claim – and we confront some of them below. But for the moment let us recall that the question before us is whether the realist can provide an internally consistent sketch of the history of science that is miracle free – not whether he can compel the anti-realist to agree the sketch is accurate. I have argued that the realist can do this.

Before proceeding, a point about the relationship between my argument for the plausibility of take-off theories such as Mendelian genetics and the high frequency assumption (the assumption that the rate at which empirically successful theories emerge is too high to be attributed to chance, cf. 4.3.1) should be made. The prospects of the miraculous endorsement argument for realism are tied to the high frequency assumption because, without this assumption, there is no 'miracle' that calls for explanation. For if the rate at which novelly successful theories emerge is sufficiently low (viz., if the proportion of novelly successful theories among the total population of emergent theories is sufficiently small), then the anti-realist can explain the novel successes of currently accepted theories by applying van Fraassen's Darwinian argument. I have assumed that the high frequency assumption is true in order to investigate the prospects of realism on this assumption – this investigation led to the problem of take-off theories. I have claimed to solve the problem of take-off theories by appealing to the use of atheoretic extra-empirical criteria which can be used to argue that a take-off theory has a reasonably high prior probability – a probability that of course can be increased by the provision of subsequent novel successes. But once the argument for take-off theories is provided, it might appear that the high frequency assumption is no longer needed. For if one accepts the argument for take-off theories, then one should accord such theories reasonably high priors irrespective of whether the high frequency assumption is true. Thus the high frequency assumption appears at this point to be dispensable. However, the relationship between the take-off argument and the high frequency assumption is deeper than it appears:

---

[19] And indeed, if one is to argue for the truth conducivity of some extra-empirical criterion, it is hard to know how else to proceed. For one cannot plausibly claim that any extra-empirical criterion can be known a priori to be truth conducive. One could attempt to ground the truth-conducivity of some extra-empirical criterion on its tendency to sanction theories that go on to prove novelly successful, but that would beg the question against the anti-realist in our current context. The only remaining option, it would seem, is to appeal to the success of that extra-empirical criterion at the level of horizontal inference.

suppose, e.g., that the high frequency assumption were not true, and novelly successful theories emerge no more rapidly than one would expect if the theory evaluation process were no better than random. This would cast doubt on the truth-conducivity of arguments for take-off theories – for if such arguments are correct one will expect the high frequency assumption to be true (because one will expect that take-off theories will be true, and thus turn out novelly successful, at a rate which coincides with the reasonably high prior assigned by the take-off argument to such theories – and the same goes for theories that are subsequently chosen because they cohere with take-off theories). Thus the high frequency assumption remains important to the overall case for realism that is developed here. Magnus and Callender (2004) argue that there are fundamental reasons why claims like the high frequency assumption – which describe the rate at which true or empirically successful theories emerge in the history of science – cannot be known, but I remain unconvinced of this. In any case, if such statistical claims are unknowable, then the argument for take-off theories that is the lynchpin of my version of the miraculous endorsement argument can stand without them.

4.3.2.2 *The anti-realist challenge*
The anti-realist challenge purports to show that the anti-realist can provide a miracle free sketch of the history of science that culminates in the astonishing empirical accuracy of many contemporary theories. The disposition of contemporary scientific method to prompt the endorsement of theories which have proven empirically successful at a rate too high to be accounted for by chance can be explained by regarding contemporary background theories K as not true but empirically adequate. Now the question how K came to be endorsed must be confronted – and the anti-realist can explain the endorsement of K by claiming that K was favored by methodological principles based on earlier background theory K' which was also (approximately) empirically adequate. But now the anti-realist faces the problem of take-off theories – how is the emergence of a first good theory to be explained – where by 'good' we now mean simply empirically adequate? The most plausible answer to this question that I believe the anti-realist can give is to mimic the realist answer given above: first good theories emerge because they provide an explanation of a body of data that meets some extra-empirical criteria that do not depend for their credibility on any coherence with background theory. In cases like the ones discussed above, the extra-empirical criterion will be some version of the simplicity criterion. But the anti-realist will naturally refuse to grant that simplicity is

truth conducive – if he does not, he is no anti-realist after all. But the anti-realist cannot in our present context simply refuse to grant the truth conducivity of simplicity. If he is to argue that the emergence of an empirically adequate take-off theory is non-miraculous, he must argue that the simplicity of the theory is good reason to believe that the theory is (and will thus subsequently prove to be) approximately empirically adequate – and his argument must not depend on any argument to the effect that simplicity is truth conducive. If simplicity is not empirical adequacy conducive, then it will simply be a miracle if the take-off theory should go on to correctly predict novel phenomena that the theory was not designed by its designer to explain. On what basis can the anti-realist argue for the empirical adequacy-conducivity of simplicity?

Let us recall the venerable realist tradition that defends the truth conducivity of simplicity at the level of vertical inference by an appeal to the apparent truth conducivity of simplicity at the level of horizontal inference. Insofar as simplicity considerations are apparently truth conducive when inferring claims about unobserved but observable entities (a claim many anti-realists would concede), why not extrapolate this truth conducivity to claims about unobservable entities? But to this van Fraassen had an ingenious reply (1980: 19–21). When it comes to some statement O about observable entities, the claim that O is true is logically equivalent to the claim that it is empirically adequate (we refer to this claim below as 'the equivalence claim'). But this means that our ordinary practices of horizontal inference admit of at least two interpretations. Competent human subjects who accept O because it is the simplest explanation of some evidence E might be interpreted as committed to a rule that sanctions the inferred truth of simplest explanations – but might just as well be interpreted as committed to a rule that sanctions the inferred empirical adequacy of simplest explanations. But this means that the ordinary practices of horizontal inference can be appealed to by the anti-realist to show that simplicity judgments are empirical adequacy conducive just as the realist can appeal to them to show that they are truth conducive. So the anti-realist will claim that the emergence of take-off theories that are empirically adequate – but not necessarily approximately true – can be explained by claiming that simplicity is conducive to empirical adequacy but not necessarily to approximate truth (and that this claim is consistent with the known facts about accepted horizontal inference). Thus the anti-realist can provide a miracle free sketch of the history of science that culminates in the astonishing empirical adequacy of contemporary theory. At this point the realist and anti-realist would seem to be on equally solid ground.

My position is that this conclusion is erroneous. Notice, first of all, that in the argument just given, the anti-realist appeals to the ordinary practices of horizontal inference to show that simplicity is conducive to empirical adequacy, but of course it is critical that the anti-realist also claim that simplicity can be conducive to empirical adequacy while not necessarily being truth conducive. But this requires that the inference to some claim's empirical adequacy can be made legitimately while avoiding any inference to the truth of the claim. Thus the anti-realist position requires that one can separate legitimate inferences to empirical adequacy from inferences to truth – I will call this the 'separability requirement.' It should be noted that the realist is not committed to the separability requirement. In judging that some theory is probably true, the realist simultaneously judges that the theory is probably empirically adequate – for truth entails empirical adequacy. But what evidence is there that the separability requirement can be met? Insofar as both the realist and anti-realist appeal to the ordinary practices of horizontal inference to support their respective positions, we should consider whether the separability requirement is met at the horizontal level. But I would argue that it is not. If we agree with van Fraassen that the equivalence claim is true (and if we assume that agents realize that it is true, as seems fair), then it is inconceivable that an agent may infer the empirical adequacy of O without inferring its truth. So the inference to empirical adequacy cannot be legitimately made without the inference to truth – so the separability requirement is not met at the horizontal level. Insofar as the realist does not require separability, but the anti-realist does, this conclusion favors the realist.

Against the argument just given, one might deny van Fraassen's equivalence claim – and thus maintain that it is at least conceivable that the separability requirement could be met at the horizontal level. Psillos, for example, argues that "strictly speaking" van Fraassen's equivalence claim does not hold – the claims that 'all observable phenomena are as-if a mouse is in the wainscoting' and 'a mouse is in the wainscoting' are not logically equivalent (1997: 371). For it might be the case, he contends, that all observable phenomena are as if a mouse is in the wainscoting without there being a mouse in the wainscoting – I might hear the pitter patter of little feet, see the alleged mouse droppings, my cat may scratch aggressively at the wainscoting – without there actually being a real mouse in the wainscoting.

Obviously the dispute between van Fraassen and Psillos over the equivalence claim is ultimately a result of their holding distinct views about what it is for some phenomenon to be 'observable.' For van Fraassen, for some entity O to be observable is for there to be some conceivable scenario in which a physiologically normal human observer could survey O with her unaided senses. Thus vastly distant stars qualify as observable for van Fraassen despite their being undetectable to human senses without a telescope – for it is the case that humans could survey such distant stars with their unaided senses if humans were suitably placed. Likewise a mouse in the wainscoting is an observable entity for van Fraassen – however cloistered the mouse's actual presence may be to human cohabitants. Let us refer to van Fraassen's conception of observability as 'in principle observability.' For Psillos' denial of the equivalence claim to hold, he must hold that according to some plausible conception of 'observable,' in principle observable entities are not actually observable. I suggest as the relevant conception what I will call 'practical observability.' For an entity to be practically observable for some agent A, it must not only be in principle observable, but the requisite conditions for observation must be within the realm of practicality for A in the particular context in which he finds himself. Thus a mouse in the wainscoting, while in principle observable, may not be practically observable – if the only way to actually observe the mouse is to rip up the interior wall in one's house, and if considerations of expense and convenience prohibit this investigation. Likewise distant stars, while in principle observable, are not practically observable if the task of placing a human observer in sufficient proximity for a naked eye observation of these stars is not within the realm of practicality. When Psillos claims that imputations of empirical adequacy to observation statements are not "technically" equivalent to imputations of truth to such statements, I suspect he had in mind the following claim: the imputation of practical empirical adequacy to O is not logically equivalent to the imputation of truth to O. This, of course, is true.

Suppose we replace the notion of in principle observability with that of practical observability and turn again to the realist/anti-realist dispute. Now the critical difference is that it is at least logically conceivable that the separability requirement is met at the horizontal level – one is logically permitted to infer the practical empirical adequacy of an observation statement without inferring its truth. But the anti-realist will now want to argue that in fact the ordinary practices of horizontal inference allow agents to infer that some observation statement O is practically empirically adequate without necessitating that such agents infer the truth of O – and thus argue that the separability requirement is met at the horizontal level. But I will argue that the ordinary practices of horizontal inference do not allow this – unless O has no untested novel consequences. Again, let the evidence be that I hear the pitter patter of little feet, apparent mouse droppings appear, and that food left out over night turns out apparently nibbled. Now one might insist that 'ordinarily' agents are entitled to infer that O ("A mouse is living in the wainscoting") is true – but there might well be actual agents – of a cautious nature characteristic of the anti-realist – who want to stop short of inferring this. A cautious agent might insist only that O be deemed practically empirically adequate – all phenomena whose observation is within the realm of practical possibility are as if O is true. O itself, according to this agent, should not be regarded as even probably true. Is the cautious agent in a tenable situation? Not, I would argue, if the agent's situation evolves so that O comes to have additional practically observable consequences that are both novel and as yet untested. Suppose it turns out that according to reputable biologists, mice always leave a characteristic substance behind wherever they walk – the substance is undetectable to unaided human senses, but can be revealed by an appropriate piece of observational technology. Now if O is true the instrument will reveal the traces – and suppose that the agent's situation evolves so that the use of this instrument is now within the realm of practicality. Now it seems clear that the cautious agent who has refrained from judging that O is true has no reason to think that the instrument will reveal the traces – and thus no reason to regard O as practically empirically adequate. For in the absence of the presumed truth of O, what reason would the cautious agent have for expecting the mouse traces to be revealed?[20] A truly cautious agent would have to abandon his cautious nature to insist that O is at this point practically empirically adequate. If the cautious agent were to learn that the mouse traces were revealed, he would have no choice but to regard it as a lucky accident. The point is that the ordinary practices of horizontal inference do not allow for inferences to the practical empirical adequacy of observation statements when such statements have as yet untested novel consequences. Thus even in this case, in which imputations of truth to observation statements are not logically equivalent to imputations of empirical adequacy, the separability requirement that the anti-realist needs is not met at the horizontal level.

---

[20] The anti-realist might respond that his reason is simply the fact that O is the simplest explanation of the observed phenomena, and that he regards simplicity as empirical adequacy conducive. But in fact the anti-realist is not entitled to this response – since he is trying to appeal to the common practices of horizontal inference to ground the claim that simplicity is empirical adequacy conducive despite not being truth conducive. Thus he must argue that ordinary cautious agents behave as if this were true – and I am arguing that they clearly do not.

This last point reiterates a common objection that realists make to anti-realists: anti-realists have no basis for regarding theories as empirically adequate when such theories have untested novel consequences. At the most general level, this objection simply amounts to the version of the miraculous endorsement argument for realism defended here – the anti-realist must regard the novel empirical successes of endorsed theories as miraculous. But the anti-realist challenge purported to show that this was not true, for the anti-realist allegedly could provide a miracle free account of the historical process that produced contemporary method (which we have assumed generates empirically adequate theories at a rate too high to be accounted for by chance). I have argued that upon scrutiny the anti-realist challenge founders – it must regard the endorsement of empirically adequate take-off theories as miraculous – and thus the proponent of the anti-realist challenge has, unlike the realist, simply 'buried the miracle' into remote regions of the history of science.

The miracle argument is often called 'the ultimate argument for realism' presumably because it is widely regarded as the most compelling and important argument for realism. Anti-realists have responded to this argument in various ways: (1) there is no need to explain the success of science, (2) the miracle argument is circular, and (3) anti-realists can nonetheless offer cogent explanations of the success of science. Anti-realism thus appears to be a position that can survive the refutation of any two of these arguments provided that the third remains. At stake in this section has been (3) which is clearly of independent interest. I have argued that the large literature on this subject suffers from a general failure to acknowledge the distinction between two versions of the miracle argument, and from a general preoccupation with just one version. In my judgement, the miraculous theory argument is not the most promising argument for the realist – it seems destined to provoke only an unending dispute about which semantic property of theories (truth or empirical adequacy) suffices to explain the success of science. The miraculous endorsement argument is, in my view, a more promising realist alternative. But let us not forget that the prospects for the realist are critically tied (in a way described above) to an assumption that has not been argued for here: namely, that the rate at which empirically successful theories emerge is significantly higher than one can attribute to chance – a claim I have deemed the high frequency assumption. If this assumption is false, the miraculous endorsement argument falls to van Fraassen's Darwinian response.

The failure of the anti-realist challenge to the miraculous endorsement argument is of considerable importance to the account of virtuous predictivism expounded in Chapter 3. Thus far virtuous predictivism has been characterized as the position that novel confirmations carry more weight than non-novel confirmations because novel confirmations testify to the truth – *or at least the empirical adequacy* – of the endorser's background beliefs. Insofar as one thinks of such background beliefs as theoretical beliefs, then, virtuous predictivism would seem to be an account of scientific method that is available to both the realist and the anti-realist. For while the realist will feel free to count novel success as evidence for the truth (or approximate truth) of such theoretical background beliefs, the anti-realist presumably would feel free to count novel success as evidence for just the empirical adequacy of theoretical background beliefs. Now the anti-realist could do this if she were able to provide a solution to the problem of take-off theories that meets the separability requirement – for with such a solution she could note that anti-realism has the resources to explain the systematic ability of scientists to identify theories that turn out empirically successful at a rate too high to attribute to chance (i.e. to explain the truth of the high frequency assumption) – which will amount to an explanation of scientists' systematic ability to endorse novelly successful theories. She would then be prepared to take novel success as evidence for the empirical adequacy (but not the truth) of background belief. But I have argued in this section that the anti-realist cannot provide such a solution – which leaves the anti-realist unable to explain the truth of the high frequency assumption, which leaves her with no systematic explanation of scientists' endorsement of novelly successful theories. Thus novel success, for the anti-realist, can provide no special evidence of empirical adequacy of endorser background beliefs. Virtuous predictivism is not available to the anti-realist. Thus virtuous predictivism should be characterized as the claim that novel success carries special weight because prediction testifies specifically to the truth (and not mere empirical adequacy) of the endorser's background beliefs.

Can the anti-realist embrace unvirtuous predictivism? I see little reason to think so. Unvirtuous predictivism holds when accommodators are prone to endorse theories that are built to fit data in some disreputable way – either because such theories violate some extra-empirical criteria or because they incorporate ad hoc hypotheses that are insufficiently supported either by extra-empirical criteria or empirical data. Thus a theory that predicts true N is preferred by an unvirtuous evaluator over one that accommodates N. But of course extra-empirical criteria, like background belief, are not regarded as truth-conducive by the anti-realist in any case – nor has the anti-realist any reason to regard them as conducive to empirical adequacy (this point follows from the discussion above – since the belief that some extra-empirical criterion is conducive to truth or empirical adequacy can simply be glossed as a background belief). Thus it seems that the anti-realist has no reason to prefer theories that respect such criteria over those that violate them with respect to the issue of the probability of a theory's truth or empirical adequacy – at most, the anti-realist will prefer theories that respect extra-empirical criteria on pragmatic grounds. This leaves the anti-realist no apparent basis to endorse predictivism either in its virtuous or unvirtuous varieties.[21]

4.4 THE BASE RATE FALLACY

P. D. Magnus and Craig Callender (hereafter, 'M&C') argue in their (2004) article that much of the literature that addresses the realist/anti-realist controversy suffers from a basic error known as the 'base rate fallacy,' and their analysis serves as a useful foil to our present project. Comparing M&C's analysis to the analysis of this chapter is made a bit more difficult because their analysis applies to what I have deemed the miraculous theory argument, rather than the miraculous endorsement argument I develop here. I will illustrate their point about the former argument, note a rather severe problem with it, and then show that their point can be applied to the miraculous endorsement argument as well and provide a satisfying rejoinder to the corresponding problem with the argument. To illustrate their point as simply as possible (and without reference to novelty) realists have tried to argue for a high probability for theories that have turned out to have true empirical consequences on the grounds that it would be 'miraculous' if false theories had true empirical consequences (the miraculous theory argument). But this argument fails to appreciate that (to put the

---

[21] The entire question of the anti-realist's stance on predictivism raises the somewhat vexing question of how the anti-realist assigns probabilities to theories in any event. As noted in footnote 18, van Fraassen argues that the constructive empiricist should not assign point-valued probabilities to theories, but only interval valued probabilities whose lower bound is 0 and whose upper bound is p(E), where E is the conjunction of all the theory's empirical consequences (1989: 194). This means that if the various propositions that constitute E are confirmed one by one, the value of p(E) rises, increasing the size of the interval. The confirmation of E serves not to increase the probability of the theory but to simply make the probability increasingly 'vague.' Perhaps the anti-realist could equate the growing vagueness of the interval with a kind of increasing theory confirmation. At most the anti-realist could view a disreputable accommodator as prone to endorse ad hoc hypotheses that are simply insufficiently supported by data – so that the procedure of conjoining a theory with such an ad hoc hypothesis produces a conjunction whose empirical consequences E* have a diminished probability – thereby having an interval valued probability with a low upper bound. This leaves to the anti-realist a rather watered-down version of unvirtuous predictivism.

point in Bayesian terms) that mere empirical success cannot argue for a high posterior probability without information about the prior probability of a theory. The point is obvious by inspection of Bayes' theorem – if h entails e, then even if p(e) is low one cannot compute p(h/e) without knowing p(h). A similar failure to acknowledge the importance of prior probabilities infects the anti-realist argument known as the 'pessimistic induction.' M&C argue that realists and anti-realists alike have cast their arguments in terms that ignore the need for prior probabilities – thus committing the base rate fallacy – and claim hereby to illuminate the sense of 'ennui' that permeates the realist/anti-realist debate.

The point can be recast in terms of the miraculous endorsement argument: the fact that a scientist (or scientific community) endorses a theory that goes on to enjoy novel success provides no basis for assigning a high probability to the background beliefs of the predictor in the absence of information about the prior probability of those background beliefs. With this point I agree, hence the fact that my analysis of predictivism in Chapter 3 incorporated specific assumptions about the prior probabilities of the predictor's and the accommodator's background beliefs (based on the presumed epistemic credibility of endorsers), and hence the fact that my preferred version of the miracle argument in this chapter incorporated an analysis of how to argue for the not-too-low prior probabilities of take off theories (by appeal to atheoretic extra-empirical considerations like simplicity which are arguably truth-conducive) – as well as an assumption about the prior probability that an emergent theory is empirically successful (this is essentially what the high frequency assumption stipulates).

M&C distinguish between what they deem 'wholesale arguments' for realism which purport to offer a sweeping justification of all currently accepted theory and 'retail arguments' for realism which purport to justify only some particular theory. Wholesale arguments, they argue, should be abandoned, for there is no way to craft them without committing the base rate fallacy. Attempts to determine the relevant base rates are thwarted by severe methodological problems of counting the theories that have been proposed and of determining with any confidence which of them are actually true. Retail arguments, however, remain possible, for it is possible in some cases to offer specific arguments for the claim that a particular theory has a reasonably high prior probability – say by appealing to currently accepted background knowledge and showing the theory to cohere with such knowledge. The realist/anti-realist debate thus should proceed on a case-by-case basis, and the general question of whether accepted theories are true should be set aside.

But there is a straightforward response to this strategy, and it is that the background theories that scientists appeal to so as to establish prior probabilities themselves require justification. M&C's appeal to presumably true background theory thus appears question-begging in the context of any defense of realism, including a retail defense, for what reason is there to think that the background theory is true? One could point to the empirical successes (or specifically novel successes) of such background theory, but the relevant prior probabilities are needed once again if the base rate fallacy is to be avoided. M&C are aware of this problem, but are not particularly worried about it. Referring to what I call 'background theory' as 'independent theory,' they write:

"One may object that these independent theories themselves require base rates, theories corroborating these base rates require base rates, and so on." In other words, maybe there is no way of breaking out of the circle involving base rates. We see no particular reason to believe that this is the case. If it is and the circle is wide enough, however, then this claim begins to sound like merely a statement of the problem of general skepticism – not something we are concerned to tackle here. (2004: 335)

It seems to me that M&C have underestimated the severity of this objection to their argument for the viability of retail realism. I find their last suggestion unsatisfactory – for it is not clear to me that the realist/anti-realist debate is not in some broad sense concerned with the problem of general skepticism. It is true that both sides to this debate typically agree that the realm of the macroscopic is knowable, contrary to general skepticism, but beyond this proviso (which is really a mere assumption for which neither side supplies an argument) the anti-realist is simply a skeptic who denies the possibility of theoretical knowledge. Thus it seems that if there 'is no way to break out of the circle' then the anti-realist wins, at least in the sense that no non-question begging argument for realism has been provided. M&C's case against wholesale realism is based in part on the unknowability of the relevant base rates – such unknowability results because it is entirely unclear how to statistically determine the frequency of true theories among the totality of 'candidate theories.' Thus they are doubtful that any version of what I have called the high frequency assumption can be established as true. It is entirely unclear how their confidence in the background theories that they need to defend retail realism can be reconciled with this general skepticism about base rates.

My response is to argue that there is a way to break out of the circle: it is by way of the epistemic history of contemporary background beliefs that trace back to take-off points – the story I sketch in (4.3.2.1) above. My example of this was the argumentation that induced Bateson to accept an early version of Mendelian genetics. Such argumentation proceeds without appeal to accepted theory but only by appeal to accepted knowledge of the observable realm. The extra-empirical criteria (such as simplicity) that the observable realm deems truth-conducive are extrapolated to serve at the theoretical level as well. Take-off theories can be assigned a reasonably high prior on the basis of the fact that they meet such criteria together with accommodated evidence. But this stops the regress, for the realist and anti-realist agree to the knowability of the observable realm. Once take-off theories are given a reasonably high prior, the realist can point to their subsequent empirical successes as evidence that their posterior is sufficient for acceptance, and such theories become established background knowledge, capable of directing subsequent theory evaluation. Once a critical mass of take-off theories is assembled, subsequent theory choice based in part on coherence with these theories should prove both truth conducive and conducive to novel success.

The argument for retail realism (in the context of the miraculous theory argument) given by M&C also seems vulnerable to the sort of objection I deemed the 'anti-realist challenge' in (4.3.2.2). They argue for reasonably high prior probabilities for candidate theories in particular scientific contexts by appealing to arguments for such theories based on background beliefs that are accepted as true on some basis or other. But why couldn't the anti-realist respond by claiming that it is no less reasonable to attribute mere empirical adequacy to such background beliefs, rather than truth? This would entail that subsequent empirical success simply increased the probability of empirical adequacy, rather than truth. The same point could be applied to the corresponding argument for retail realism in the context of the miraculous endorsement argument: rather than explain the success of scientists in endorsing novelly successful theories by attributing truth to their background beliefs, why not simply attribute empirical adequacy to such background beliefs? But this sort of objection was squarely addressed above in terms of the inability of the anti-realist to establish that scientific method underwrites the conclusion that take-off theories should be regarded as empirically adequate but not true. Novel success consequently cannot count in the anti-realist's thinking as evidence of the endorser's empirically adequate background beliefs.

I have defended in this chapter a particular version of the miracle argument for realism because I believe that it is viable and it is the version that incorporates a coherent understanding of the epistemic significance of novelty. But more important for present purposes is the fact explained in the prior section: anti-realism is inconsistent with virtuous predictivism and, for the most part, unvirtuous predictivism as well.

4.5 THE REALIST IS A PLURALIST

I want to conclude this chapter by returning to the dichotomy between epistemic individualism and pluralism and considering how it bears on the realist/anti-realist debate. On the one hand, the idea that the proponent of scientific realism is a pluralist seems rather plausible. For the proverbial proponent of scientific realism sounds somewhat like the non-expert evaluator of (2.4.1), who points to the empirical, and especially the novel, successes of science as evidence that scientific theory is very credible, and probably approximately true. Now of course if the realist simply points to the empirical successes of science as evidence of the truth of theory, the realist has committed the base rate fallacy, for he has ignored the need for well grounded prior probabilities. The realist could attempt to master all the relevant sciences, of course, and thus come to know which background beliefs prevail in those sciences, and thus come to appreciate the arguments for emerging theories from such background beliefs that serve in some cases to establish certain theories as having reasonably high priors. But for the would-be proponent of wholesale realism, this is probably a too gargantuan task. Such a person could, like other non-experts, take refuge in epistemic pluralism, holding that the intellectual credibility of the mature scientific communities in question offers some reason to believe that the requisite background knowledge exists and does offer reason for reasonably high priors – and one need not possess such knowledge oneself to know this. Thus pluralism might be thought to offer hope for a base-rate respecting wholesale realism.

While some may find this line of analysis appealing, others will surely not. For wholesale realism will strike some philosophers as by its very nature too broad a generalization to be plausible. Such philosophers will, if they are inclined to realism, prefer retail realism, with its detailed appeals to specific scientific arguments of the sort scientists themselves consider persuasive. But the retail realist should not imagine that she has embraced a rugged epistemic individualism. For we saw in Chapters 2 and 3 that expert scientists themselves often must resort to a pluralistic method in their assessments of theories. Such expert pluralists included interdisciplinary, imperfect, humble and reflective expert evaluators (cf. (2.4.2) and (3.8.3)). There is no reason to think that the proponent of retail realism will be in any better position to disregard the epistemic significance of other experts' judgments. The appeal to local background theory of the sort recommended by M&C, for example, will be permissible only for those who have reason to regard such theory as well grounded, and this may well require trusting the judgments of an earlier generation of scientists who accepted it. The realist – whether a scientist or philosopher – will find it hard to avoid epistemic pluralism.

The anti-realist, by contrast, seems to have epistemic individualism as an option. In fact, a strong commitment to individualism may drive one to anti-realism – for the simple reason that as an individualist one is unwilling to take others' judgments as epistemically significant. But, as we have seen, realism depends heavily on such willingness. The anti-realist thus proves herself to be a less trusting person, but this is hardly surprising: she is, after all, a skeptic.

CHAPTER 5

The predicting community

5.1 INTRODUCTION

Jeane Dixon was one of the best known astrologers and alleged psychics of the twentieth century. Probably her most famous prediction was the assassination of John F. Kennedy. But there is a well known explanation for Dixon's ability to make astonishing correct predictions, and it is the fact that she made many more predictions that were unsuccessful but largely ignored. If one makes enough predictions, some of them are bound to turn out true by chance. The mathematician John Allen Paulos defined the "Jeane Dixon effect" as the tendency to tout successful predictions while ignoring unsuccessful ones. Clearly, the epistemic import of a successful prediction has something to do with the number of predictions on offer.

It was noted in the previous chapter that the pool of endorsements in which an endorsement of a novelly successful theory is made can affect the epistemic significance of that endorsement. The high frequency assumption claims that the rate at which novelly successful theories are endorsed is too great to be accounted for by chance – for if it were lower than this, there would be little reason to take any particular endorsement of a novelly successful theory as a result of the endorser's holding true background beliefs. Such endorsements could simply be the result of chance, i.e., false background beliefs that happened to prompt an endorsement of a theory with true novel consequences. To put the point another way, for any fixed number of endorsements of novelly successful theories, the degree to which such endorsements legitimate an inference of true background beliefs diminishes as the pool of endorsements in which they occur enlarges. In the last chapter I argued that only the realist can provide an explanation (other than chance) for the truth of the high frequency assumption.

This sort of point can apply as well to theory evaluation at a local level. To return to the recurring thought experiment of this book: when one purchases the services of a financial advisor, for example, one essentially trusts that the predictions of that advisor will prove a good investment over time. It would seem to make sense, therefore, to hire an advisor who has a good predictive track record. However, one's trust in an advisor with a strong predictive record could be reduced if that advisor were one of a large pool of advisors – if the pool is large enough, and if one located the successful advisor by systematically searching through the entire pool, then the fact that that advisor performed well could easily be the result of chance. Evidence that financial advisors in general have some ability to pre-select high performing investments could come in the form of information about the ratio of high performing advisors to all advisors – a claim which roughly corresponds to the high frequency assumption of the previous chapter.

In this chapter I will explore these issues in more detail. My strategy will be to make use of Maher's coin flip example. This example, while obviously idealized, is wonderfully simple and instructive, and lends itself nicely to our current project. I summarize the example in the next section, but I do make various changes to suit my own purposes (for Maher's own version, see (1.4.5)). I then go on to provide an interpretation based on my own theory of predictivism. I then develop the example by considering variations in the size of the prediction pool and the number of successful predictors located within it. Ultimately the analysis here will serve to enrich our grasp of predictivism by revealing it to be a social phenomenon requiring a social level of analysis.

5.2   THE COIN FLIP EXAMPLE ONCE AGAIN

As noted in Chapter 1, Maher's (1988) first considers a subject – the predictor – who predicts the outcome of 100 coin flips. We will here refer to this predicted sequence as E+. Thereafter the coin is flipped 99 times, and each flip results in just the predicted outcome – this conjunction of the apparently random 99 outcomes is deemed E. I will suppose that the predictor predicts E+ because he has endorsed[1] a theory which entails E+.[2]

---

[1] References in this chapter to acts of endorsement should be construed as postings of endorsement-level probabilities.

[2] Maher's notation is different – he uses 'T' to refer to what I call 'E+,' and does not stipulate that the predictor makes his prediction T because he is committed to an underlying theory that entails that prediction (though Maher does apply his account of predictivism to this sort of case in his (1993)). I believe the version I give here is more faithful to a typical act of prediction, one in which a predicted observation statement is based on an underlying theory (rather than a conjunction of observation

In what follows, however, it will simplify matters if we focus not on the evaluation of this theory per se but on the evaluation of E+. This squares with the idea that, at least in some cases, we are more interested in the accuracy of the predictions than in the truth of the underlying theory (as in the case of the mutual fund investor). We assume, for simplicity, a non-expert evaluator (one with no views of her own about the dynamics of coin flips) who would presumably assign high probability to E+ on this basis.

In a different scenario, another subject – the accommodator – is initially presented with E. On that basis he endorses some theory that entails E+, thus the accommodator predicts E+ (more precisely, he predicts the last conjunct in E+). We are now to consider what probability a reasonable pluralist evaluator (not herself a coin flipologist) would assign to E+. Clearly, the only reasonable reply is that the theory involved is substantially less probable for a reasonable evaluator in the accommodator's scenario than T is in the predictor's scenario – despite the fact that it would seem to be just evidence E that is offered in support of E+ in each case. This is a clear illustration of the predictivist intuition.

Maher's explanation of the predictivist intuition in this case is that the successful predictions in the predictor's scenario constitute persuasive evidence that the predictor 'has a reliable method' of making predictions – and that no such evidence is available in the accommodator's scenario. This could be interpreted to mean that the predictor has a reliable method of constructing a theory. I discussed this interpretation in (1.4.5) and offered various criticisms of it in (3.8.4.1). My own explanation is along the lines of the account of predictivism developed in Chapter 3: the fact that the predictor endorses E+ suggests his possession of independent evidence (which we refer to as the predictor's 'background beliefs') and the subsequent demonstration of E constitutes substantial evidence for the truth of the predictor's background assumptions – whereas no such evidence is available in the accommodator's scenario. More specifically, when we are confronted with the predictor's success in the first scenario, we are faced with the following dilemma once E is observed: either the predictor has true (or, if one prefers, approximately true) background beliefs or has false (or not approximately true) background beliefs but has luckily endorsed a truth entailing theory nonetheless. (I assume for simplicity that any

statements that entail that prediction). Of course, Maher's example also describes predictors and accommodators as constructers of theories, rather than endorsers, given his commitment to use-novelty.

predictor with true background beliefs will endorse a true theory and thus make true predictions.) But if we judge that the prior probability of a randomly selected theorist having true background beliefs is (though perhaps small) much greater than the probability that a randomly selected theorist has false background beliefs but nonetheless will happen to endorse an E-entailing theory (without appeal to knowledge of E), then we will judge that the predictor probably has true background beliefs. So what ultimately drives the inference that the predictor's background beliefs are true is our antecedent judgment that

$$(\%)\,P(K) \gg P(K[E]/\sim K)P(\sim K)$$

where K refers to the background beliefs of the predictor and K[E] asserts that K prompts the endorsement of an E-entailing theory (without appeal to E). 'K' occurs not as a rigid designator but as a non-rigid term that refers to the background beliefs of whatever theorist is at stake. Probability function P takes as background information a particular view of the relative frequency of true background beliefs in the predicting community (this fixes the value of $P(K)$), and also the knowledge of E. Function P also reflects a certain logical non-omniscience on the part of the evaluator, for the logical consequences of the various K are not established (hence $P(K[E])$ can take values other than 0 or 1). This non-omniscience reflects the presumed pluralism of the evaluator – who does not know the actual content of K (and hence does not know its consequences) and/or relies on information provided by the endorser for knowledge of its consequences – and relies on the degree of trust she places in the endorser for her assigned prior probability to K. For example, suppose $P(K) = 0.05$ while $P(K[E]/\sim K)P(\sim K) = 0.001$; this entails that an E-predictor is, given the truth of E, 50 times more likely to have true background beliefs than not (for it is fifty times more likely that an arbitrary predictor's background beliefs are true – and thus will prompt an E-prediction if E is true – than that an arbitrary predictor will happen to predict E though his background beliefs are false). But in that any predictor's background beliefs must be either true or false, it follows that the probability of a successful E-predictor's background beliefs being true is 0.98.[3]

On the other hand, since the accommodator did not predict E, we are not faced with the dilemma of explaining a successful prediction, hence are

[3] For – assuming E is shown true – let z be the probability that K is false but happened to predict E, and 50z be the probability that K is true. Then $z + 50z = 1$; hence $50z = 0.98$.

free to point out that the probability of the accommodator's predictive method being reliable is just the value of P(K) (assumed above to be 0.05). Either the accommodator is virtuous, in which case his endorsement probability accurately reflects the probability he should assign to E+, given his background beliefs, or he is unvirtuous and has somehow cooked his theory to fit E. But in either case, we – who are for the moment playing the role of evaluators – are left with no basis for an increased confidence in the accommodator's background beliefs – in the case of the virtuous accommodator, there is no success predictivism to drive up the evaluator's confidence in K (for the accommodator's prior probability for E+ was, given my conception of 'accommodator,' below level L), and in the case of the unvirtuous accommodator, there is no reason to trust that E+ has any strong evidential basis at all.

5.3 EVA MEETS PRISCILLA

Imagine that Eva, our pluralist evaluator, is presented with E (the initial 99 flip outcomes) and considers now what probability should be assigned to E+ (which conjoins E with a prediction that the 100th flip outcome will be heads). Insofar as she has no reason to think that E+ is true rather than E' (which conjoins E with a prediction that the 100th flip will be tails), she assigns probability 0.5 to E+, for she knows the coin is fair. Eva now encounters Priscilla the predictor, who informs Eva that she endorsed a theory which entails E+ prior to seeing the outcomes constituting E and thus successfully predicted E. On the basis of this information, and assuming relevant prior probabilities about Priscilla's background beliefs, Eva assigns a probability near one to E+.

Let us now consider Eva in isolation again. We assume she does not know whether an E+-predictor like Priscilla exists or not, so Eva still assigns probability 0.5 to E+. Now it would seem reasonable at this point for Eva to concede that if she were to learn of an E+-predictor's existence (i.e. learn that there was a predictor who predicted E+ in advance of E's becoming known) she should raise her estimate of E+'s probability to near one (assuming she were not to acquire any other evidence in favor of or against E+ itself). We deem this counterfactual claim '(A)':

(A) If Eva (who now knows that E) were to learn that an E+-predictor existed prior to the demonstration of E, Eva should raise her estimate of E+'s probability to near one (we assume Eva does not acquire any other evidence for or against E+).

It would seem to be in accordance with the account of predictivism sketched in Chapter 3, and other accounts as well, to accept (A). But while (A) has intuitive appeal, it falls apart on closer inspection. The failing of (A) is its assumption that the epistemic significance for the rational degree of belief in E+ of the E+-predictor's existence can be evaluated independently of various facts about the relevant portion of the scientific community of which the E+-predictor is a member. In the remainder of this chapter we will consider various ways in which this proves true. For the sake of simplicity and concreteness we will continue to think in terms of Maher's coin flip example.

5.4 THE COMMUNITY OF PREDICTING SCIENTISTS

We consider now the portion of the scientific community that has been working on the same problem that Eva is considering: the identification of the true empirical hypothesis describing the 100 coin flip outcomes of our example. More precisely, let us consider the portion of this community that made predictions regarding the outcomes of the 100 flips prior to the first coin flip – we deem this 'the predicting community' for this empirical domain. Now suppose that Eva (who is not a member of this community as she did not make a prediction before witnessing the first 99 flips) learns that there are exactly N-many coin flipologists in this predicting community, thus N-many predictions of the flip outcomes were made (some of which may be identical – we assume only one prediction per predictor). Now the question becomes what is the epistemic significance for E+ of the supposition that there was an E+-predictor (i.e. at least one) amongst the predicting community prior to the establishment of E. The relevant point here is that this epistemic significance is surely not independent of the number N. For the larger is N, the greater the probability that there will be at least one E+-predictor within the N-many predicting coin flipologists even if E+ is false, for the probability becomes greater that some predictor with false background beliefs will happen to endorse an E+-entailing theory. This shows that (A) is too simple to be an adequate characterization of the epistemic significance for E+ of an E+-predictor's existence.

We recall Eva's position: she witnesses E (the outcome of the first 99 flips) and then considers what probability might be assigned to E+ – since there is no reason to think that the 100th flip will be a heads, and given her belief that the coin is fair, we stipulate that, for Eva, P(E+) = 0.5. Let us define 'D' as 'There is at least one E+-predictor in the predicting community.' The problem is to define $P(E+/D)$ for arbitrary N. For some N Bayes' theorem of course provides that:

$$P(E+/D) = \frac{P(E+)P(D/E+)}{P(D/E+)P(E+) + P(D/\sim E+)P(\sim E+)} \tag{1}$$

We recall that for Eva, $P(E+) = P(\sim E+) = 0.5$. So (1) reduces in this context immediately to:

$$P(E+/D) = \frac{P(D/E+)}{P(D/E+) + P(D/\sim E+)} \tag{2}$$

Now $P(D/E+)$ will be equal to the probability that – assuming E+ – not all of the N-many predicting scientists fail to predict E+. We can determine $P(D/E+)$ thus by subtracting the probability that all fail to predict E+ (assuming E+) from 1 (i.e., $P(D/E+) = 1 - P(\sim D/E+)$). Now the prior probability that an arbitrary predictor with background beliefs K will happen to predict E+ (assuming E+) is $P(K[E+]/E+)$; the prior probability that an arbitrary predictor will predict some outcome other than E+ (assuming E+) is $P(K[\sim E+]/E+)$. We assume the probability that one predictor will fail to predict E+ is independent of the probability of what any other arbitrary predictor will predict, so the probability that all N-many predictors will predict some outcome other than E+ (assuming E+) is $P(K([\sim E+]/E+)^N$. Hence, the probability that at least one predictor will predict E+ assuming E+ is $1 - P(K[\sim E+]/E+)^N$. Substituting in (2) we derive:

$$P(E+/D) = \frac{1 - P(K[\sim E+]/E+)^N}{\left(1 - P(K[\sim E+]/E+)^N\right) + \left(1 - P(K[\sim E]/\sim E+)^N\right)} \tag{3}$$

Let us pause here to convince ourselves of the following fact: if $N = 1$ so that there is only one member of the predicting community, $P(E+/D)$ is very close to 1. However, as N increases, $P(E+/D)$ drops toward the original prior probability of E+, 0.5. The intuition in back of this result is straightforward: if the predicting community contains only an E+-predictor, and E is confirmed, it is very likely (for reasons summarized above) that the predictor's background beliefs are true, thus very likely that E+ is true. But as the size of the predicting community gets larger, the fact that there is at least one E+-predictor gets less and less significant for E+. For increasing N it becomes more probable that some predictor with false background beliefs will endorse an E+-entailing theory – thus the degree of new confirmation provided by the assurance of an E+-predictor's existence falls off as N increases (though the probability of E+ should not drop below the original value of 0.5, as the information that at least one E+-predictor exists should never disconfirm E+).

To convince ourselves of this, let us plug some plausible values into (3) and plot P(E+/D) against N. The terms in need of values are P(K[~E+]/E+) = 1 − P(K[E+]/E+) and P(K[~E+]/~E+) = 1 − P(K[E+]/~E+) (these equalities hold since we assume each K prompts the endorsement of some theory which entails some values for all the relevant coin flip outcomes), so let us stipulate plausible values for P(K[E+]/E+) and P(K[E+]/~E+). Our assumption (%) that P(K) ≫ P(K[E]/~K)P(~K) asserts that it is much more likely that an arbitrary set of background beliefs (in the relevant community) is true than that a set of background beliefs will prompt the endorsement of an E-entailing theory while nonetheless being false. This assumption, however, entails that P(K[E+]/E+) ≫ P(K[E+]/~E+) for arbitrary background belief K – for if E+ is true then E+ is much more likely to be predicted by an arbitrary method than if E+ is false. This is because if E+ is true then it will be predicted by any background beliefs which are either true, on the one hand, or false but nonetheless E+ entailing – but if E+ is false then E+ can only be predicted by false E+ entailing background beliefs. But again, (%) entails that it is much more likely that an arbitrary set of background beliefs are true than that they are false but E-entailing (and recall that E is 99/100's of E+), so P(K[E+]/E+) ≫ P(K[E+]/~E+).

Let us pause at this point to take note of a small technical point. If we stick to Maher's original example in which E+ contains one hundred conjuncts, then P(K[E+]/~E+) will be somewhat less than the probability that a random heads/tales generator will generate a particular sequence of 100 outcomes – but this probability will be the mind-bogglingly small $(0.5)^{100}$, or 7.89×10^{-31}. (Only false K can generate E+ if E+ is false, and the vast majority of Ks are false. I assume the probability of false K generating a theory that gets any particular coin flip outcome right is just 0.5.) This number is, I think, too small to mirror real world analogues of the coin flip example and will prove computationally messy in the calculations to follow. I propose at this point to redefine E+ to consist of an apparently random sequence of 10 (not 100) flip outcomes (the last outcome being heads); E is thus reconstrued as the initial apparently random 9 outcomes. Below we make assumptions that respect this modified version of the coin flip example. We will work with this version of the example through the rest of this chapter.

$$\text{Assumptions: } P(K[E+]/E+) = 0.05$$
$$P(K[E+]/{\sim}E+) = 0.001^4 \tag{4}$$

So, $P(K[{\sim}E+]/E+) = 0.95$ and $P(K[{\sim}E+]/{\sim}E+) = 0.999$ (given our assumption that any set of background beliefs will generate some prediction of coin flip outcomes). Substituting into (3) we obtain:

$$P(E+/D) = \frac{1 - (0.95)^N}{(1 - 0.95^N) + (1 - 0.999^N)} \tag{5}$$

So for $N = 1$, $P(E+/D) = 0.98$; $P(E+/D)$ clearly drops toward 0.5 for increasing N, as Figure 5.1 shows. [The data points are as follows: for $N = 1$: $P(E+/D) = 0.98$; $N = 500$: $P(E+/D) = 0.72$; 1000: 0.61; 1500: 0.56; 2000 = 0.54; 2500 = 0.52; 3000 = 0.51.)

Now let us vary the information Eva receives in the following way: instead of learning that there is at least one E+-predictor among the N-many predicting scientists, let us assume that she is informed that there is exactly one E+-predictor in the predicting community. What should her updated probability for E+ be given this information? Intuitively, for sufficiently small N the existence of a unique E+-predictor should still count as strong evidence for E+ once E is established. But for

---

[4] Actually, $P(K[E+]/{\sim}E+)$ will be approximately equal to 0.000927, but we round to 0.001 for computational ease, and because this does not distort any subsequent outcome. The stipulation that $P(K[E+]/E+) = 0.05$ is made to respect the assumption that $P(K[E+]/E+) \gg P(K[E+]/{\sim}E+)$ but is otherwise arbitrary. Consider:

$$P(K[E+]/E+) = 0.05 = P(K[E+]/E + K)P(K)$$
$$+ P(K[E+]/E+ {\sim}K)[1 - P(K)] \tag{a}$$

But $P(K[E+]/E+ K) = 1$, as any true set of background beliefs will generate a true (and thus truth-entailing) theory (this is one of our assumptions), and $P(K[E+]/E+ {\sim}K)$ is essentially equal to the probability that a sequence of fair coin flips will produce a particular sequence of 10 heads/tails outcomes, which is 0.5^{10} or 0.000976. Inserting these values in (a) and rearranging gives $P(K) = 0.049072$. Now,

$$P(K[E+]/{\sim}E+) = P(K[E+]/{\sim}E + K)P(K) + P(K[E+]/{\sim}E+ {\sim}K)$$
$$[1 - P(K)] \tag{b}$$

But of course $P(K[E+]/{\sim}E+ K) = 0$, as true background beliefs never generate falsehood-endorsing theories. Furthermore $P(K[E+]/{\sim}E+ {\sim}K) = P(K[E+]/E+ {\sim}K) = 0.000976$, since if a set of background beliefs is false it is just as likely to predict E+ whether E+ or not E+ is true; using the above value for $P(K)$ entails $P(K[E+]/{\sim}E+)$ is 0.000927.

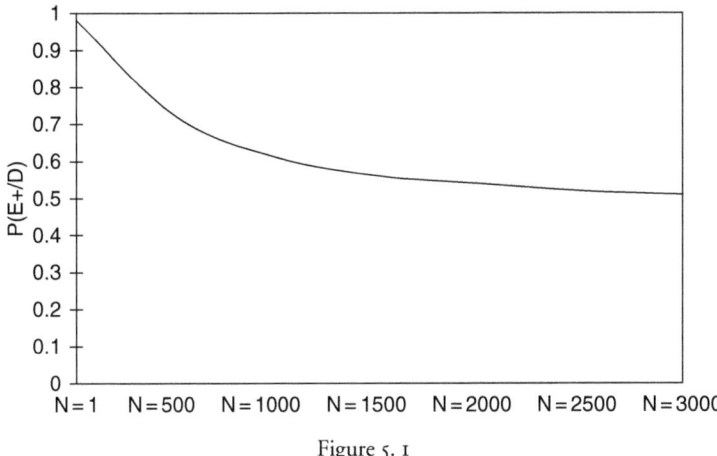

Figure 5. 1

growing N, the existence of but one predictor will increasingly count against E+! This is primarily because if E+ is true, all predictors with true background beliefs will predict E+ – but then if E+ is true and there is only one E+ predictor, this entails that there is at most one predictor with true background beliefs in the community, a claim increasingly improbable for sufficiently large N (given our assumption that P(K) is not too low). If E+ is false, the existence of a unique predictor is more understandable for large N, as a unique E+-predictor is compatible in this event with there being any number of predictors with true background beliefs in this community (assuming at least one predictor has false background beliefs).

Let 'D!' assert that the predicting community contains exactly one E+-predictor. We seek the value of P(E+/D!) for arbitrary N. Again, given Eva's situation, P(E+) = P(~E+) = 0.5 (recall she has witnessed E by this point), so Bayes' theorem gives

$$P(E+/D!) = \frac{P(D!/E+)}{P(D!/E+) + P(D!/\sim E+)} \qquad (5)$$

Assuming a predicting community of N-many members, any scenario in which some particular predictor is the unique E+-predictor and the remaining N − 1 predictors predict some outcome sequence other than E+ will have a probability equal to $P(K[E+]/E+)[P(K[\sim E+]/E+)]^{N-1}$. In that there are N-many possible states in which a unique predictor exists (corresponding to the possibility that each of the N-predicting scientists is the unique E+-predictor), the probability that there is a unique E+-predictor if E+ is true is $N[P(K[E+]/E+)P(K[\sim E+]/E+)^{N-1}]$. By identical reasoning, the probability that there is a unique E+-predictor if E+ is false is just $N[P(K[E+]/\sim E+)P(K[\sim E+]/\sim E+)^{N-1}]$. So, substituting in (5), and canceling the N's, we have

$$P(E+/D!) = \frac{P(K[E+]/E+)[P(K[\sim E+]/E+)^{N-1}}{P(K[E+]/E+)[P(K[\sim E+]/E+)]^{N-1} + P(K[E+]/\sim E+)[P(K[\sim E+]/\sim E+)]^{N-1}}$$

$$(6)$$

To show how P(E+/D!) varies with increasing N let us plug in some appropriate numbers and plot the conditional probability against N. We use the same assumptions as given in (4):

$$P(T/D!) = \frac{(0.05)(0.95)^{N-1}}{0.05(.95)^{N-1} + 0.001(0.999)^{N-1}} \qquad (7)$$

In Figure 5.2 we represent P(E+/D!) against rising N. (The data points are N = 1: P(E+/D!) = 0.98; 50: 0.81, 100: 0.256, and 150: 0.027.)

The point to appreciate is that the information that there is a unique T-predictor can have widely varying epistemic import for T depending on the size of N – while the case of N = 1 is indistinguishable from the case of N = 1 where there is at least one E+-predictor, P(E+/D!) falls off astonishingly rapidly for increasing N (give our assumptions as described in (4)). For if E+ were true there would almost certainly be more than one

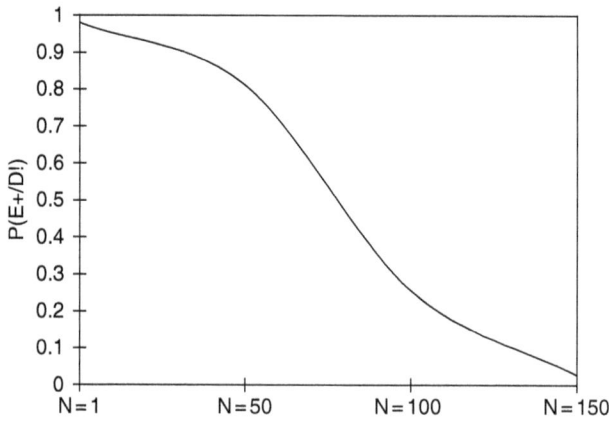

Figure 5. 2

E+-predictor on our assumptions – the absence of additional E+-predictors thus counts against E+ for large N.[5]

Let us consider again Eva. We recall that we have modified our assumptions about her evidence and theory; she has witnessed E (the initial, apparently random, sequence of nine coin flip outcomes) and considered the hypothesis E+ that results from conjoining E with the hypothesis that the 10th flip will turn up heads; she has at this point attached the probability of 0.5 to E+ (she assumes the coin is fair). Now let us suppose that Eva randomly encounters the E+-predictor Priscilla, and by reasoning explained above Eva raises her degree of belief in E+ to near 1. Now let us suppose that Eva subsequently learns that Priscilla was part of a vast predicting community for the coin flip problem – in fact she was one of 3000 predictors for this problem! Now it might seem as though Eva should lower her degree of belief in E+ to 0.51 (see Figure 5.1), given that P(E+/D) = 0.51 for N = 3000. But this is terribly counterintuitive – why should it matter with respect to Priscilla's putative reliability that there happened to be 2999 other predictors pondering the same problem?

In fact, it doesn't matter, and Eva should, given the way the above scenario was described, continue to attach high probability to E+. The reason is that Eva encountered Priscilla 'at random,' viz., as a result of what we imagine to be something like a single random sampling of the predicting community. This being so, it is nonetheless vastly more likely that Eva is a reliable predictor than that she is unreliable but happened to predict T nonetheless, as explained above, hence very likely that E+ is true. However, suppose that Eva discovered Priscilla's existence not by a random encounter but as the result of a systematic search through all 3000 predictors for an E+-predictor. In this event, Eva's discovery of Priscilla amounts to nothing more than the information that there is at least one

---

[5] Readers may indeed be astonished at the steep slope of the graph in Figure 5.2 as compared with Figure 5.1 – the reason for the difference can be found in our assumptions made in (4), which establish that if E+ is true, about 5 out of a hundred predictors should endorse the true E+-entailing theory but if E+ is false, only about 1 out of 1000 should do so. Thus for, say N = 200, there will probably be about 10 E+-predictors if E+ is true but most likely none if E+ is false. So the information that 'at least one E+-predictor exists' will tend to be tightly correlated with the existence of other E+-predictors – and thus confirm E+ rather highly for this community size. But if we are assured that ONLY one E+-predictor exists out of 200, this will make E+'s probably exceedingly low, for this information blocks the inference from the existence of the one E+-predictor to the existence of others.

E+-predictor among 3000, and Eva should simply set her degree of belief in E+ on D equal to P(E+/D) for N = 3000 (0.51). The reason for the epistemic asymmetry between Eva's random and non-random sampling methods derives from the assumptions we made in (4): if the sampling method is a single random selection, it is extremely unlikely that the method will produce an E+-predictor if E+ is false (given that P(K[E+]/~E+) is so low). However, if E+ is true, the probability of it producing an E+-predictor is comparatively much greater (since P(K[E+]/E+) ≫ P(K[E+]/~E+)). Since the random selection of an E+-predictor is much less surprising assuming E+ rather than ~E+, it thereby confirms E+. But if the sampling method is exhaustive by design and thus non-random, then the method will reveal any E+-predictor's existence including those whose background beliefs are false no matter how few there are of these in the predicting community. Hence it turns out that the epistemic significance of an E+-predictor's existence depends not only on how large the predicting community is, but also on the method by which an E+-predictor's existence is revealed to the evaluator (the example of the mutual funds in (5.1) can be used to illustrate this point).

Let us consider another problem: imagine Eva again, prior to learning of any E+-predictor's existence, still with degree of belief 0.5 in E+. As before Eva encounters Priscilla at random – and learning of Priscilla that she is an E+-predictor prompts Eva to raise her degree of belief in E+ to near one. Now in this case, Priscilla is one of a member of 3000 predictors for the coin flip problem, but also informed that in fact Priscilla is the unique E+-predictor in this community. Now, extrapolating from the analysis presented in Figure 5.2, it looks as though Eva should lower her degree of belief to a tiny probability, as P(E+/D!) will clearly be extremely low (Figure 5.2 gives P(E+/D!) as 0.027 for N = 150; for N = 3000 the value will be vastly lower). But should it matter vis à vis Priscilla's apparent reliability that she was a member of a community of 3000 predictors, one in which she was the only E+-predictor? The answer, surprisingly, is yes! For by analysis presented above, the failure of other predictors to predict E+ constitutes strong evidence against E+ for sufficiently large N, and the realization that there are no other E+-predictors counts as new and powerful evidence against E+ for such N. The data graphed in Figure 5.2 is of course based on the assumed values for P(K[E+]/E+) and P(K[E+]/~K) given in (4) – but the same basic point will hold for any non-zero values assumed for these conditional probability such that P(K[E+]/E+) > P(K[E+]/~E+); though difference assumptions may shift the position and slope of the graph, it will remain a downward curve drifting toward 0 for increasing N.

The reason, again, that the realization that Priscilla is a unique E+-predictor in a community of 3000 forces Eva to lower her degree of belief in E+, while the realization that Priscilla was simply one of possibly more E+-predictors in the community does not, is that the latter realization, unlike the former, is compatible with there being any number of other E+-predictors in the community – thus the realization does not constitute evidence against E+, while the former does.

5.6 COUNTERPREDICTORS

Let us return to Eva one last time, and imagine that she randomly encounters Priscilla the E+-predictor, whereupon Eva raises her degree of belief in E+ from 0.5 to near 1 once E is established. But now let us suppose that Eva randomly encounters another predictor, who is addressed as 'Countess.' Now in advance of the demonstration of E Countess predicted E', where E' asserts E in combination with the prediction that the 10th flip would be tails; E' thus agrees with E+ on the initial nine predictions (which constitute E) but differs on the last flip prediction. The point, of course, is that upon the demonstration of E Countess has demonstrated no less predictive skill than Priscilla, and thus E counts equally in favor of E+ and E'. Countess is an example of what I deem a 'counterpredictor' with respect to the E+ predictor Priscilla; she is a predictor of a set of observable outcomes inconsistent with E+ but which, like E+, accords with E.

What is the probability of E+ conditional on the existence of an E+-predictor and a E'-predictor once E is known? Once E is known $P(E+ \vee E') = 1$, but the total evidence here clearly favors neither outcome sequence over the other, so of course the probability of E+ in this case is 0.5. For the same reason E+ will deserve the same conditional probability if there is an equal number of E+-predictors and counterpredictors no matter what this number is (a proof of this claim is analogous to the proof provided below).

Suppose that Eva happens to encounter Priscilla and Countess, but then happens to meet another E+-predictor. What is the updated probability of E+ on the existence of two E+-predictors but just one E'-predictor? Upon meeting Priscilla and Countess, Eva's new probability for E+ is 0.5; this situation is epistemically equivalent (as regards the probability of E+) to one in which Eva has not encountered any predictors. After meeting the second E+-predictor, Eva should presumably be free to regard this E+-predictor's existence as confirming E+ just as strongly as if Eva had not met the first two predictors, for the counteracting predictions of Priscilla and

Countess do not apparently undermine the significance of the second E+-predictor's existence, so it seems her probability for E+ in this event should be near one.

Let's attempt to sharpen the above analysis. We consider three randomly encountered members of the predicting community we deem A, B, and C. A and B are E+-predictors (which we symbolize as A+ and B +) while C is an E'-predictor (C'). We seek the value of P(E+/A + B+C') in a context in which each predictor successfully predicted E. Bayes' theorem gives:

$$P(E+/A + B + C') = \frac{P(E+)P(A + B + C'/E+)}{P(A + B + C')} \qquad (8)$$

Since the predictors' predictive behaviors are mutually independent, we have

$$P(A + B + C'/E+) = [P(A+/E+)P(B+/E+)]P(C'/E+) \qquad (9)$$

Let's consider the three conditional probabilities on the right hand side of (9). P(A +/E+), for example, is the probability that predictor A will predict E+ on the assumption that E+ is true – this is clearly intended to be the probability that A will predict a true outcome sequence given the fact that A has demonstrated considerable predictive skill over the initial nine predictions. We must therefore be careful in determining which information we include in the background knowledge on the basis of which our conditional probabilities are calculated, for if we include the fact that A has predicted E+, P(A +/E+) will have value 1, a value which does not reflect the intended probability. Let us therefore stipulate that the background knowledge on whose basis a probability function P' is selected includes only the fact that A, B, and C have generated their initial nine predictions which collectively constitute E (predictions on which they all agree) and that E has been shown true. Now P'(A +/E+) is equal to the probability that A has true background beliefs (and hence will surely predict E+ if E+ is true) plus the probability that A has false background beliefs but will nonetheless happen to endorse an E+-entailing theory – where both probabilities take account of the fact that A has successfully predicted E. We let Ka refer to the background beliefs held by A:

$$P'(A + /E+) = P'(A + /E + Ka)P'(Ka) + P'(A + /E+ \sim Ka)(1 - P'(Ka)) \qquad (10)$$

P'(A +/E+ Ka) is 1, since any agent with true background beliefs will predict the true outcome sequence; P'(A +/E+ ∼Ka) is 0.5, since if A's background beliefs are false we should assume that it is as likely that A will predict heads for the 10th flip as tails. We thus require only a value for P'(Ka) (the probability that a set of background beliefs that have successfully predicted E are true). Surprisingly, we have come this far in our analysis without fixing the value of this probability. The only constraints on this value in our current context are the pair of assumptions made in (4). These assumptions entail that P'(Ka) = 0.96.[6] Now the probability that Ka will prompt an E+-prediction on the assumption that E+ is true is as follows:

$$P'(A+/E+) = P'(A+/E + Ka)P'(Ka) + P'(A+/E+ \sim Ka)[1 - P'(Ka)]$$
$$= (1)(0.96) + (0.5)(0.04)$$
$$= 0.98. \tag{11}$$

By identical reasoning, P'(B+ /E+) = 0.98 as well. Now P'(C+ /E+) = 0.98 as well, so where P'(C$\sim$+ /E+) is the probability that C will predict some outcome sequence other than E+ (assuming E+), P'(C$\sim$+ /E+) = 0.02 (since C will certainly predict either E+ or some outcome sequence other than E+). Given that C has already predicted E, C will predict E' if and only if C predicts some outcome sequence other than E+, hence P'(C$\sim$+ /E+) = P'(C'/E+) = 0.02. Substituting into (9),

$$P'(A + B + C'/E+) = [(0.98)0.98]0.02$$
$$= 0.019208 \tag{12}$$

Now of course

$$P'(A + B + C') = P'(A + B + C'/E+)P'(E+)$$
$$+ P'(A + B + C'/\sim E+)[1 - P'(E+)]. \tag{13}$$

---

[6] Consider the probability function P* that takes only K[E] (not E) as background knowledge (where K is any set of background beliefs that have generated the prediction of E, such as that held by A, B, or C); now we have

$$P^*(K/E) = \frac{P^*(K)P^*(E/K)}{P^*(E/K)P^*(K) + P^*(E/\sim K)[1 - P^*(K))]}$$

But P*(E/K) = 1, given K[E] as background, and P*(K) = 0.049 (the computations in footnote 4 that show P(K) = 0.049 are identical to the computations that show P*(K) = 0.049, as the presence of background knowledge of E (used in computing probability function P) does not affect any relevant values). P*(E/$\sim$K) is the probability (given no other relevant information) that E will happen to turn out true, which is $(0.5)^9$ or 0.00195. This gives P*(K/E) = 0.9635. But P'(K) = P*(K/E) by definition of P'.

But again, given the presumed knowledge of E, we know ∼E+ is equivalent to E', so

$$P'(A + B + C') = P'(A + B + C'/E+)P'(E+)$$
$$+ P'(A + B + C'/E')[1 - P'(E+)]. \quad (14)$$

Now by reasoning structurally similar to that running from (9) to (12), we know that

$$P'(A + B + C'/E') = [P'(A + /E')P(B + /E')]P'(C'/E')$$
$$= [0.02(0.02)]0.98 \quad (15)$$
$$= 0.000392$$

So, substituting into (14), we have

$$P'(A + B + C') = (0.019208)(0.5) + (0.000392)(0.5) \quad (16)$$
$$= 0.0098$$

Substituting now into (8),

$$P'(E + /A+B + C') = \frac{0.5(0.019208)}{0.0098} \quad (17)$$
$$= 0.98.$$

Thus the probability of E+ on the condition that A and B have predicted E+, while C has predicted E', is identical to the probability of E+ conditioned only on the existence of a single E+-predictor. This result squares with the intuition explained above: the E'-predictor together with one of the E+-predictors mutually annihilate each other's epistemic significance regarding E+, but the remaining E+-predictor counts no less impressively for E+. In fact, it is a straightforward generalization of the reasoning applied above to show that the probability of E+ on the random encounter of x-many E+-predictors and x − w E'-predictors (given w ≤ x) is identical to the probability of E+ on x − w E+-predictors where no E'-predictors have been sighted.[7]

---

[7] It may well seem surprising to some that, given our Assumptions in (4), the existence of, say, 1001 E+-predictors and 1000 E'-predictors should be epistemically equivalent for our rational degree of belief in E+ to a scenario in which a single randomly encountered E+-predictor exists, but this is precisely the case. The temptation to think otherwise is based, I suspect, on the fact that if there really were 1001 E+-predictors and 1000 E'-predictors prior to the demonstration of E, we would be strongly inclined to deny the principle on which the assumptions in (4) were based – specifically to deny the principle (%) that $P(K) \gg P(K[E]/\sim K)[1 - P(K)]$. For the almost even distribution of E+ and E'-predictors shows that among those predictors who endorse some theory that entails E, it is not

5.7 APPLICATIONS

It goes without saying that the results reached below are highly idealized. What requires illumination is the way in which these results apply to the actual history and methodology of science. Consider the following: Immanuel Velikovsky concocted a bizarre theory of the history of the solar system (Velikovsky 1950, 1953) which, he argued, had the virtue of explaining the remarkable amount of similarity between the stories of various ancient religious and mythological texts. The scientific community universally denounced Velikovsky's theories, but it was nonetheless granted that he managed to make some surprisingly successful predictions – and there were no clear falsifications of his theory (owing in part to Velikovsky's willingness to adopt ad hoc hypotheses when problems threatened). Contrary to then prevailing opinion, Velikovsky successfully predicted that Venus would be hot, that Jupiter would emit radio waves, and that the earth has a magnetosphere. The willingness of the community to discount these successful predictions can be explained in part on the basis of the community's judgment that his arguments were patently unsound. But the following point is surely true as well: if we are to widen the definition of 'scientist' so far as to include the likes of Velikovsky, we will thereby include in the scientific community just about anyone who ever cared to make a bold prediction based on some kind of evidence. It is scarcely possible that none of this vast community of predictors will ever be lucky enough to watch their bold predictions come true. The willingness of the scientific community to denounce such occasionally successful prognosticators as Jeane Dixon and Nostradamus can be straightforwardly explained in terms of the social dimension of predictivism, whatever other explanations we might adduce as well.

One reason it is difficult to find clear historical illustrations of the epistemic relevance of the size of the community of predictors is that the actual number of scientists trying to construct theories of any particular empirical domain is typically quite small. Scientific research is a highly specialized affair, not only because successful research requires much training but because scientists have less incentive to tackle problems that a large number of other scientists are already working on, since the chance of

the case that many more of them have true background beliefs than false background beliefs (because, out of 2001 E-predictors, at least 1000 have false background beliefs) – which in turn implies that (%) is probably false. If we abandon (%) then of course we lose our motivation for the current result. If we retain (4) then the 1001/1000 E+/E'-predictors result is amazingly unlikely, but nonetheless would establish the probability of E+ to be 0.98.

succeeding where others fail is smaller when the playing field is large. So the epistemic relevance of the size of the predicting community has a tendency to remain invisible in the history of science. This invisibility, however, should not diminish the philosopher's interest in the social nature of predictivism, insofar as it is her aim to construct a complete theory of confirmation. And we saw in the previous chapter how the relative frequency of novelly successful predictions determines the truth or falsehood of the high frequency assumption, a point intimately related to the realist/ anti-realist debate.

The actual relevance of the size of the predicting community is played out in part in the determination of what degree of novelty is required for a prediction to count as sufficiently bold. For example, the fact that there are known to be relatively few scientists attempting to construct theories (and thus make novel predictions) pertaining to some empirical domain entails that predictions need not be so bold to count as successful demonstrations as they would need to be if there were a huge number of scientists constructing more theories of the domain. The fact that no one knows precisely how many predictors are at work in a particular domain need not affect this point – the rougher the estimate of the number, the rougher the corresponding determination of the required level of predictive boldness for an impressive experimental demonstration of a theory.

Does the fact that there is only one predictor of some theory among a large predicting community ever actually work to discredit the theory? It surely does. Suppose that among the very large community of economists who attempt to predict the degree of growth in the American economy next year only one predicts that there will be no growth – all others predict growth of some degree. At the end of the first quarter of the next year the lone economist's bold prediction is surprisingly vindicated – no growth is observed! Nonetheless, the fact that none of the other economists (some of whom must be the most knowledgeable alive) endorsed the hypothesis that there would be zero growth for the entire year works to accord low probability to this hypothesis despite the lone economist's successful prediction – growth will almost surely pick up in the remaining quarters.

The above results also find application to an issue discussed by Kahn, Landsberg and Stockman (1992) (whose theory of predictivism was discussed in (1.4.5) and (3.8.4.3). KLS present a simple social planner's problem.

Imagine a planner who would like to build a bridge, and is seeking a scientific theory to guide its design. If the theory is true, the bridge will stand and if the theory is false the bridge will fail. The planner can direct the activities of a fixed population of scientists, and can require them all to either look first [i.e. acquire data and accommodate it in theory construction] or theorize first [endorse a theory and then test its predictions]. What should he do? (1992: 511)

KLS go on to pose the following solution to the problem:

If there are many researchers, each working independently and each theorizing first, then the probability that all of their theories will be rejected is very small. Therefore, since the planner requires only one surviving theory, he should order researchers to theorize first. This increases the probability that the bridge will be built and stand, at the cost of only a very small risk that no bridge can be built at all. (1992: 512)

While the proposed solution is tempting, it reveals a clear failure to appreciate the relationship between the size of the predicting community and the epistemic significance of the existence of a successful predictor. KLS seem unaware that a large predicting community is naturally more likely than a smaller one to contain a scientist who endorses a theory that entails an outcome sequence that will survive rigorous testing but is nonetheless false (like an E'-predictor if E+ turns out true). A failure to appreciate the social nature of predictivism is not merely a theoretical failure – it could have disastrous consequences for bridge walkers!

5.8 REALISM, PLURALISM, AND INDIVIDUALISM

At the onset of the thought experiment described in this chapter, Eva was described as a pluralist. Her pluralism was evident from her willingness to accept as evidence facts about the endorsements of scientists of a particular theory – more precisely, about the number of scientists making such endorsements (and others). This reveals a biographicalism characteristic of the pluralist. Insofar as Eva is prepared to assign a modestly high probability to the truth of the background beliefs of an arbitrary predictor, moreover, Eva reveals herself to be a realist. For it is only the realist who views human agents as having a systematic ability to identify true (or approximately true) theories of the world, including background theories. This squares with a fundamental result of the previous chapter – for were Eva to be an anti-realist, she could not be a predictivist. Were Eva an anti-realist, she would view all novel success – whether derived from true or false background beliefs – as the result of chance. This is because her prior probability that an arbitrary predictor has true background beliefs would be sufficiently low that the possession of true background beliefs is primarily a matter of luck.

What if Eva is an individualist? As such she denies epistemic significance to acts of endorsement and surveys the various background beliefs held by the various members of the predicting community, as well as the evidence that supports them, entirely by herself. As an individualist there is less reason to ascribe the sort of logical non-omniscience to her that we ascribed to pluralist Eva, who depends on endorsers for knowledge of logical consequences in a way individualist Eva refuses to. Given the individualist's detailed knowledge of the various K at issue and the totality of relevant evidence, she will assign prior probabilities to the various background beliefs and the theories they support – as evidence comes in, she will simply update those beliefs by Bayesian conditionalization. Thus the analysis provided above does not in any obvious way apply to her – and the tempered socialized predictivism that held of the pluralist will not either.

But it will be recalled that there are two versions of weak predictivism, tempered and thin. As an individualist Eva cannot be a tempered predictivist – but the question remains whether a thin form of predictivism could characterize her deliberations if she were presented with scenarios like those described in this chapter, viz., predictors who are members of predicting communities of varying sizes. This would be the case if the size of the predicting community, while not directly epistemically relevant, were linked with some other factor F that the individualist did perceive as relevant.

One important distinction between individualist and pluralist in our present context is that the individualist will not be as directly focused on the size of the predicting community per se. The pluralist focused on this factor because of her presumption that a large predicting community has a larger probability of containing a predictor whose background beliefs led to novelly successful but false theories. This reflects the pluralist's assumption that the number of predictors either does or at least may roughly correspond to the number of different sets of background beliefs that are in play. However, a growing number of predictors will not increase the probability that some predictor will get predictions right by chance if it is known that all the predictors possess the same background beliefs. While it was allowed in the above analysis that some predictors may possess the same background beliefs (indeed, for the realist a non-negligible group will have true, and hence presumably identical, background beliefs) it was nonetheless important for the pluralist that it is at least possible that any particular predictor has a different set of background beliefs than any other – hence the size of the predicting community is seen as correlated with the probability that some predictor will get predictions right by chance. But while the pluralist focuses on the size of the predicting community, the individualist will look directly at the various background beliefs in play. Now the question is whether the individualist will discern a correlation between a large number of competing background beliefs and a diminished posterior probability for any particular theory, sanctioned by one set of background beliefs. This strikes me as a reasonable possibility for at least two reasons. One is that a scientific field that is characterized by a very large number of competing background beliefs may contain a fair number of candidate background beliefs that are just silly – and will be accorded a low prior by the individualist. This low prior will generate a low posterior even if the resulting theory is novelly successful. Another possibility is that, even in the absence of silly background beliefs, a field characterized by a large number of competing (and all at least minimally plausible) background beliefs will perforce assign to each of the competition a low prior, for the very reason that there are so many live options to be considered. Once again a low posterior for novelly successful theories will result. This amounts to grounds for a thin social predictivism for individualist evaluators.

5.9 CONCLUSION

Maher argued at the conclusion of his (1988) that the truth of predictivism had remained hard to see for so long because so many philosophers were convinced that facts about the method by which a theory is constructed could have no relevance to the degree of confirmation a theory enjoyed on the available evidence – Maher's method-based theory of predictivism (according to him) belies this conviction. I disagree with this assessment, of course, because I deny the epistemic significance of construction. My conceptualization of predictivism preserves the distinction between the contexts of discovery and justification that Maher's repudiates. I would argue, however, that the truth of predictivism (in its tempered form) has remained hard to see because so many philosophers are epistemic individualists who are inclined to deny the epistemic relevance of acts of endorsement. I would argue analogously that the social dimension of predictivism has proved likewise hard to see because there is a widespread assumption among philosophers of a Bayesian ilk that has blocked our view: that assumption is that social factors will not prove to have a substantive bearing on the nature of the evidence relation, whatever their role in determining the actual course of scientific development. We have found reason in this chapter to deny that assumption.

CHAPTER 6

Back to epistemic pluralism

6.1 INTRODUCTION

I argued in Chapter 2 that there is a long tradition in Western epistemology of epistemic individualism. While this is true, epistemic pluralism has always had its proponents as well. In Plato's dialogue Laches, for example, the character of Socrates argues that it is important to care what people think about what one does, so long as the people whose opinions are valued are credible authorities on the matter at hand. If one's goal is athletic excellence, then, the opinion of the competent trainer is of great importance. I have explicated the notion of an agent's epistemic authority in terms of the concept of the agent's background beliefs.

The background beliefs of endorsers and evaluators have played a central role in the account of predictivism I have developed in the previous three chapters. I have been deliberately vague about what kinds of beliefs may constitute background beliefs. An agent's beliefs, according to me, are simply all the propositions to which the agent (explicitly or implicitly) assigns probabilities that are very near one. Thus any theoretical, methodological, or metaphysical claim which was accorded such a probability by an agent would count as a belief – my presumption is that when an agent assigns a probability to some claim that falls short of being near one, the agent typically has as a basis for such a probability a set of background beliefs. A background belief, in a particular scenario, is simply a belief other than the beliefs that are salient in that scenario (such as T, O, or N, if any of these are believed) but is nonetheless epistemically relevant in that scenario. Background beliefs include any (relevant) statement that expresses some fact that was established by straightforward observation – any so-called observation statement.[1] I imagine background beliefs as an inclusive and heterogeneous mixture of high probability statements.

---

[1] I have throughout this book assumed the viability of some version of the theory/observation distinction. The theory/observation distinction amounts to the claim while some facts about the world can be straightforwardly established by direct looking ('observation'), others can be established

As I have explained at some length, there are two forms of tempered predictivism: virtuous predictivism (in which the evaluator regards the posted probabilities of endorsers as logically impeccable) and unvirtuous predictivism (in which the evaluators regard such postings as logically suspect). As was noted in (3.8.3), however, a pluralist evaluator may lack full knowledge of an endorser's epistemic basis, and thus may not know whether the endorser's logic is correct or not. Thus the evaluator may well be uncertain whether the endorsers she faces are virtuous or unvirtuous. I will refer to an evaluator who is uncertain about the virtue of endorsers as an 'agnostic evaluator.' It was noted that since both virtue and unvirtue are associated with a preference for prediction, agnosticism is compatible with a preference for prediction.

But there is another important point about agnostic predictivism that it is time to make, and that is that novel success can function as evidence for endorser virtue for agnostic evaluators. Imagine an agnostic evaluator Agnes faces a predictor Paul who endorses some theory on a basis that does not include N, and N is subsequently shown true. Because of her agnosticism, it seems that Agnes cannot take Paul's endorsement (viz., his 'prediction per se') as by itself carrying much epistemic weight, for she is unconvinced that Paul is virtuous. Now given that per se predictivism does not (apparently) characterize Agnes's theory evaluation, it would seem that success predictivism cannot hold either – for success predictivism works by confirming whatever background beliefs prompted the initial endorsement – but there is little basis for presuming that these background beliefs exist. Thus virtuous predictivism, it seems, should fail. But this description of the situation is oversimple: the fact that Agnes is agnostic about Paul's posted probabilities does not entail that she believes his initial endorsement was based on distorted logic and a lack of suitable background belief – it simply means she regards the latter as possible. She could, despite her agnosticism, take the prior probability of Paul's being virtuous to be not too low (thus a glimmer of per se predictivism remains). Given that Paul endorsed the

(if at all) only indirectly by way of a process of constructing and testing theories. A simple version of the theory/observation distinction was essential to logical positivism – though one of the reasons for positivism's downfall was the realization that the distinction was an elusive one. All observation, it was and is often claimed, is theory-laden to some degree. But while the distinction has proved hard to render in a rigorous way, some version of it remains essential to much contemporary philosophy of science. The realist/anti-realist debate, which shows no signs of abating despite the frequently heard claim that it is dead, typically proceeds by appeal to some form of the theory/observation distinction. Bayesianism, for that matter, seems committed to the distinction as well, for the classical form of Bayesian learning involves conditionalization on some proposition which is established 'by observation.' It is not hard to understand why the theory/observation distinction remains a popular one – it is simply obvious that it corresponds to something fundamental and pervasive in scientific reasoning.

theory, its subsequent novel success confirms – to some degree – the claim that Paul is virtuous: he probably had background beliefs that were sufficiently comprehensive to prompt his endorsement of the theory without appeal to N.

So for all the details of the last three chapters, the account of predictivism I develop there is oversimple in one critical respect: it ignores the fact that an agent's predictive success may be evidence for something more than just the truth or truthlikeness of the agent's background beliefs. It may be evidence for what I will call the comprehensiveness of those beliefs. A belief set is 'comprehensive' with respect to some theory T insofar as that set is rich in informative content that bears on the probability of T. An endorser's predictive successes may be important because they reveal that his background beliefs are comprehensive enough to qualify him as a virtuous predictor – if we were antecedently unsure how much background information the endorser possessed and thus unsure whether the endorser was virtuous.

For example: imagine Peter the predictor and Alex the accommodator (our old friends from Chapter 3) construed this time as anthropologists. Peter is a seasoned anthropologist who has studied the Nazika people for decades – Alex is a beginning anthropologist who has only just begun to study the Nazika and has, appropriately enough, very few beliefs about them. Now the Nazika have recently undergone a change in their cultural and economic environment owing to the recent forces of modernization that have been thrust upon them by their neighbors. Prior to witnessing the effects of this modernization, Peter predicts that the Nazika will undergo a set of novel reactions N – he predicts N on the basis of T, which is a theory of how the Nazika react to novelties in their environment. He endorses T because it coheres with an established body of evidence together with Kp, his background beliefs, which are constituted by his vast body of knowledge and experience of Nazika history, culture, psychology, etc. In a different scenario, Alex simply witnesses the Nazika undergo the changes N over time, and comes up with a theory T' (which may or may not be the same as T) which he endorses primarily on the basis of N. His background beliefs (Ka) about the Nazika are true but very few and insubstantial. Agnostic evaluators (who lack this information about Kp and Ka), given only the information that Peter has successfully predicted N, may infer from this predictive success not only that Peter's background beliefs are true, but that they are also highly comprehensive, if it seems to them a lot of general knowledge would be necessary to make such predictions successfully. Agnostic evaluators, given only the information that Alex has endorsed T' on the basis of prior knowledge of N, are given little evidence of the comprehensiveness of Alex's background beliefs (or their truth, for that matter).

Another illustration: consider a history professor who has taught her students about the French Revolution. When she gives an exam over the material she typically cannot ask them about every point she taught them – she thus poses questions whose answers can be deduced from the points covered in class. Essentially such a test puts students in the position of having to 'predict' which answers will elicit a high grade on the test. Successful predictors are given a high grade because the ability to make a set of such predictions testifies to a large body of relevant (and correct) background belief, viz., the beliefs acquired through studying the material from the class. The degree of successful prediction (i.e. the percentage of questions answered correctly) probably reflects a corresponding degree in the amount of acquired background belief. Unsuccessful predictors may fail not because their background beliefs are false but because such beliefs are few in number, i.e., insufficiently comprehensive. (Students who carry a cheat sheet into the exam are 'accommodators.')

The net effect of this realization is that we must sophisticate the account of predictivism provided thus far in this book to incorporate an account of agnostic predictivism. Roughly put, the idea that predictivism is true because novel success confirms the background beliefs of the predictor corresponds to the claim that true (or truthlike) background belief suffices, given certain prior probabilities, to explain novel success, i.e., to explain the endorsement of novelly successful theories. To explain novel success is to render such success non-lucky. Thus far we have held that such success is non-lucky if the relevant background beliefs are true – but it turns out that mere truth is not sufficient for non-lucky true predictions. Background beliefs must be both true and sufficiently comprehensive to serve as a guarantor of non-lucky novel success. When agnostic predictivism holds true, consequently, it is because novel success points to background beliefs which have both features.

A set of background beliefs K may of course meet the truth requirement without meeting the comprehensiveness requirement – in the case of an unvirtuous endorser whose background beliefs happen to be true but insufficiently substantial to prompt his endorsement of T. K may also meet the comprehensiveness requirement without meeting the truth requirement: an agent may have a rich set of beliefs which, though false, nonetheless legitimate an endorsement level probability for a theory. But clearly both requirements must be met if novel success is to be rendered non-coincidental. To mark this clarification I will refer to a set of background beliefs that are, in some context, both true and suitably comprehensive as 'sufficient' in that context. Agnostic predictivism thus holds true insofar as novel success offers stronger evidence than non-novel success of the sufficiency of background belief (because the prior probability of an agent's background beliefs being sufficient is greater than the prior probability that an agent's background beliefs are insufficient but conducive to the endorsement of a novelly successful theory). The theory of predictivism spelled out in the previous three chapters should be modified to incorporate this account of agnostic predictivism.[2]

The general question whether predictivism is true is typically regarded as a question about some theory T and the evidence that confirms it. Thus the question whether predictivism held true in the case of Mendeleev's periodic law is a question about that theory and whether its novel successes carried more weight in its favor than its (otherwise similar) non-novel successes. But it is possible that in some cases the question of the theory and how it was confirmed may be less important than the issue of the background beliefs themselves. The fact that some theory enjoyed some novel success, in other words, may be important primarily for the reason that it confirmed the background beliefs of the predictor. This could be the case, e.g., when the novel success of a theory testifies that the general metaphysical views of the predictor are probably true, when those metaphysical views are deeply inconsistent with prevailing metaphysical views – thus the effect of novel success may be to engender the kind of thing Kuhn described as a scientific revolution.

Predictivism could thus be described as a two-sided thesis: on the one hand, it is a thesis about theories and how they are confirmed, on the other, it is a thesis about background beliefs and how they are assessed for truth and comprehensiveness. In line with the bulk of literature that addresses the epistemics of novel confirmation, the emphasis in the previous three chapters has been on the theory side. In this chapter we turn to the other side. More specifically, let us suppose that we are considering an epistemic context in which the full effect of some prior novel success has been fully assimilated – and the background beliefs of the predictor have been established to have some probability of being correct and their degree of

[2] For example – in the account of predictivism developed in Chapter 3 it was assumed, for simplicity, that if an agent has true background beliefs then he will endorse a true theory. But of course the point being made here is that mere truth is not sufficient to guarantee such reliability – it is sufficiency that is required. Evaluators' judgments about the prior probability that endorsers have true background beliefs should ultimately be replaced with judgments about prior probabilities of sufficient background beliefs.

comprehensiveness roughly determined. What happens next? I propose to consider an agent's competence in some field as largely a function of the truthlikeness and comprehensiveness of that agent's relevant background beliefs. How (if at all) do rational evaluators make use of information about the judgments of agents with a certain established degree of competence? This is of course not a new question for us: it was the central question of Chapter 2, which argued for the pervasiveness and legitimacy of epistemic pluralism in theory evaluation. Chapter 2 served as a primer for the account of predictivism developed in the last three chapters, for the most important form of predictivism (tempered predictivism) is a by-product of such pluralism. The account of pluralism presented in Chapter 2 was however a very informal one. In this chapter I attempt to give a more rigorous account of epistemic pluralism.

6.2 TOWARD A MORE RIGOROUS MODEL OF PLURALISTIC THEORY EVALUATION

The purpose of this chapter is to assess one natural proposal relevant to these issues. This proposal runs as follows: each agent i of X-many posters of probabilities for theory T is assigned a 'weight' w_i which reflects that agent's competence relative to the other X-1 agents (the weights sum to one). Where p_i is the probability i assigns to T, we compute the communal probability P(T) as the weighted sum of the various posted probabilities:

$$P(T) = w_1p_1 + w_2p_2 + \ldots + w_xp_x. \tag{1}$$

This technique of weighted averaging of probabilities thus provides a straightforward method of assimilating the epistemic significance of a distribution of probabilities across an epistemic community. It is a straightforward model of epistemic pluralism in action.

I find the weighted averaging method interesting because it is both *prima facie* plausible yet not obviously sound. Most importantly, it is unclear how the relevant weights are to be determined. It is not intuitively clear that a scientist's competence – understood now as including a blend of truthlikeness and comprehensiveness of background belief – is the kind of thing that can be measured by a simple numeral. Even assuming a quantitative comparison of competence, it is hardly obvious that the best way to take account of measured competence in assessing the probability of some theory is to compute the product of competence with the probability endorsed by the quantized agent and use the product as a component of an updated probability.

In the refined account of predictivism I sketched above, novel success can function as evidence of the sufficiency of the agent's background beliefs vis à vis that novel success. Thus at this moment an agent's beliefs are thought of as falling simply into one of two categories: they are either sufficient for that novel success or insufficient. But of course we must distinguish the question whether an agent's background beliefs were sufficient for some particular novel success from the question about the general competence of the agent. The judgment that an agent's background beliefs were probably sufficient for a particular novel success is a judgment that the agent probably possesses a particular chunk of background knowledge – one that was needed for that particular novel success. The question about an agent's general competence is a question about the totality of the agent's background knowledge vis à vis the field or subfield as a whole. An agent could possess sufficient background knowledge to make a (non-lucky) successful novel prediction, if such background knowledge entailed an endorsement-level probability for some truth theory T that entailed true N, while nonetheless failing to possess a lot of overall background knowledge pertinent to the field. Suppose, for example, a neophyte chemist has mastered a small portion of the field of chemistry but on this basis has managed some dramatic predictive success nonetheless. Having accorded her background beliefs as they pertained to T high probability, we should refrain from according similar high probability to whatever background beliefs bear on the next theory she endorses, for these relevant background beliefs may not be entirely the same as those that bore on T. As an agent compiles novel successes across a variety of endorsements, however, our estimate of the comprehensiveness of her overall chemical background beliefs grows accordingly. This judgment about the overall comprehensiveness of her knowledge could be relevant to the question what probability to assign the background beliefs that pertain to some subsequent theory endorsement of hers. This is not to say that novel success is the only sort of evidence of background comprehensiveness, for other sorts of evidence could well be relevant (such as the assessment of the agent's knowledge by competent evaluators, the agent's ability to handle difficult questions about her field, her educational background, etc.)

My strategy will be to consider two examples in which we isolate a particular species of competence and consider how weights might reflect it. (I then consider how to calculate 'mixed weights' which are based on estimated mixtures of both species of competence.) To describe the difference between the two examples let us first say that an agent i's relevant background beliefs Ki is a conjunction of Kei and Kti – Kei refers to the conjunction of relevant observation statements that i has come to accept as true on the basis of some straightforward basis of observation, while Kti refers to the conjunction of all other ('theoretical') statements that the agent accepts as true. Applying the theory/observation distinction to Ki, in other words, generates the bifurcation of Ki into Kei and Kti.[3] In the first scenario, individuals all possess the same theoretical beliefs but independently explore a large and diffuse body of evidence – thus their various observation-based background beliefs differ in content and comprehensiveness. A communal probability is computed by assigning weights in accordance with such differences in evidence possession. In the second scenario, a group of individuals possess a single body of evidence, but compute probabilities on the basis of varying theoretical beliefs. Weights are thus chosen to reflect the relative merits (in a sense to be explained) of such theoretical beliefs. In a final case, we consider a technique for computing 'mixed weights' which reflect differences between both evidence possession and the merit of theoretical belief. As we will see, weighted averaging turns out to be a tool that will deliver rational probability updates under various carefully described conditions.[4]

6.3 SCENARIO ONE: THE MAPMAKERS

In this scenario we consider a method of fixing weights that evaluates an agent's competence in terms of the amount of relevant evidence that agent possesses in his or her body of background belief – as noted above, this is one species of 'competence.' Let us consider a group of X-many mapmakers who have been set loose in a large complex foreign city. The city is divided into many subregions of equal area. For each mapmaker, it has been established how many subregions that mapmaker has mapped. This fact could have been established by inferring the comprehensiveness of the agent's observational knowledge from the agent's past predictive successes

---

[3] Of course agent i will presumably have background beliefs that mix theoretical and observational vocabulary, such as if T then O, which fall in neither Kei nor Kti.

[4] Before proceeding, we should note that Keith Lehrer has in a series of published works (Lehrer (1974, 1976, 1977, 1978)) made use of the weighted averaging method; the formal aspects of Lehrer's program have been formally developed by Carl Wagner (1978, 1981). Their program is laid out in detail in Lehrer and Wagner's (1981) book *Rational Consensus in Science and Society* (referred to as 'L&W' hereafter). L&W's program focuses on the use of weighted averaging to explain the phenomenon of consensus in epistemic communities. Our concern, however, is not with the generation of consensus – the primary aim of this chapter is to understand the task of weighting and to determine whether certain weighting procedures enable the averaging method to deliver rational probability updates. L&W devote only cursory attention to the problem of how weights are to be fixed, and I take note of some of what they suggest on this subject at appropriate points below.

as suggested above – or perhaps it is known independently how much time each agent has devoted to studying the city, etc. We assume that each mapmaker i has mapped H_i-many subregions of the city; no mapmaker succeeds in mapping the whole city. Since these mapmakers are merely 'observing' the city, we assume the partial maps they produce are invariably correct.

At the end of their mapmaking activity, the mapmakers are all shown a single map M, and asked to determine the probability that it is a true map of the city. One issue is whether we should imagine that the mapmakers share their evidence with one another. Initially I will assume no collaboration, i.e., each mapmaker i posts a probability p_i that M is a true map based on just the evidence that i has acquired. It might be objected at this point that the failure of our mapmakers to share their data amounts to a disanalogy with the actual practice of scientists, but as we have seen this is not clearly the case. First of all, data are often not shared between scientists for practical reasons (there is no easy way to transfer them) or social reasons (having to do with the competitive nature of science). Yet more relevant is the fact that when a scientist possesses a large amount of relevant evidence, evidence sharing may take prolonged time and effort on both sides, hence the premium on scientific expertise, which allows deference to another scientist's opinion without having to assimilate all that scientist's evidence (cf. 2.4). Below, however, I will consider a variation on the mapmaking scenario that allows for evidence sharing. Our question is whether weighted averaging should be applied to the p_i given our current assumptions – and if so, how? We assume, for the moment, that M is a true map of the city (despite this being a remarkable coincidence, given its prior probability as described below).

Let us pause to consider how individual mapmakers would determine their original posted probabilities p_i. One point of considering this scenario is to consider a case in which the various mapmakers start from the same epistemic starting point – i.e. with the same background beliefs other than the evidence garnered by their mapmaking activities. In current terms, this amounts to their shared commitment to a body of 'theoretical' background belief Kt. Kt, we will assume, amounts to the information that map M has been drawn on the basis of new satellite technology which is known to have a 0.9 probability of producing a correct map of any particular subregion. Thus Kt entails, for each mapmaker, that the prior probability of any particular subregion being correctly drawn on M is 0.9. Assuming that the probability that any particular region is accurate is independent of that of any other region, the prior probability that M is a true map is

$(0.9)^{100}$, or 2.66×10^{-5}. Each mapmaker i can thus compute $p_i = p(M/e_iKt)$ by simply using Bayes' theorem (where e_i refers to the total evidence garnered by i):

$$p_i = \frac{p(M/Kt)p(e_i/M\&Kt)}{P(e_i/Kt)} \qquad (2)$$

Thus, noting that $p(e_i/MKt)$ will of course be 1 – since e_i will accurately portray a section of M, and we are now assuming that M is correct. Hence,

$$p_i = \frac{(2.66 \times 10^{-5})}{(0.9)^{Hi}.} \qquad (3)$$

So, if H_j is 50, $p_j = 0.005$; for $H_k = 75$, $p_k = 0.072$; for $H_l = 90$, $p_l = 0.35$; for $h_Z = 99$, $p_z = 0.90$. As we would expect, p_i is closer to 1 the more subregions observed by i, exactly what we would expect given that M is a true map.

Let us attempt to apply weighted averaging to this example. The obvious question is how the various weights are to be fixed. Our mapmakers are indistinguishable from one another except for the number of subregions they have mapped – so it would seem to make sense to make each weight w_i proportional to the number of subregions i mapped vis à vis others (here, let us recall, weights aim to measure competence in one particular sense: an agent is competent to assess the probability of T to the degree that the agent is in possession of relevant evidence). The sum $H_1 + H_2 + \ldots + H_x$ gives the total number of acts of subregion-mappings performed by the community – thus let us declare $w_i = H_i/(H_1 + H_2 + \ldots + H_x)$. The basic intuition in the back of our weighting procedure is that weights reflect the amount of (reliably) garnered evidence by the weighted agent in comparison to the total amount of evidence garnered by all agents. Weighted averaging determines the probability that M is a true map – P(M) – in the way of equation (1). The question for us is whether P(M) is a rational updated probability for M given all the information we possess.

The question whether P(M) is a rational probability given the various p_i and w_i amounts to the question whether P(M) is a probability which is grounded on the total available evidence. The total available evidence in this case amounts to the totality of the city's subregions mapped by our mapmakers. Does P(M) better reflect the total evidence than, say, any particular p_i? Let us first consider a case in which a total of ten mapmakers each map ten regions – but that, as luck would have it, they happen to map precisely the same ten subregions. Thus each mapmaker i is given the weight $w_i = 0.1$, and each mapmaker posts 0.0000762. The weighted average $P(M)$ will thus turn out to be 0.0000762 – precisely the probability justified by the total evidence, since it was just the correspondence of ten subregions to map M. But now consider the following variation on this example: in this case, each of ten mapmakers maps ten subregions, but no subregion is mapped by more than one mapmaker. Thus all 100 subregions of the city are covered, and our total evidence raises the probability that M is true to 1. Unfortunately, given that the weights and the various probabilities are the same in this case as before, the weighted average determines the probability of M to be 0.0000762 – and weighted averaging clearly goes badly wrong.

We are noting at this point a fundamental problem with weighted averaging as a technique for computing probabilities. If there is a large body of evidence supporting T, but the evidence is thinly dispersed over a large number of agents who thus each post a low probability for T, weighted averaging will deliver a value that is too low. Let us consider a modified proposal: instead of requiring individual agents to post probabilities for the truth of the entire map M, we require only that they post the probability that the group of subregions they surveyed are truly represented on M – we deem this the 'modest approach' to weighted averaging. The term reflects a certain modesty on the part of probability posters, who refrain from posting probabilities about a theory about which they have too little evidence – they post probabilities only for the segment of the theory about which they have extensive expertise. Thus, assuming that any agent will correctly determine that any subregion they have mapped corresponds to M or not, every agent will post either 1 or 0 for this probability when applying the modest approach. We compute weights as before, so as to be proportionate to the amount of garnered evidence for the weighted agent. We retain for the moment our assumption that M is a true map. Does weighted averaging deliver a rational update now on the modest approach? Let us consider again the case in which each of ten mapmakers surveys ten subregions, and no subregion is covered by more than one, so that the entire city is covered. Now we have $w_i = 0.1$ for each agent as before, but each $p_i = 1$, so $P(M)$ is determined by weighted averaging to be equal to 1, precisely the correct result. So far so good – our modest approach allows the epistemic significance of disjoint evidence to be combined.

But clearly the modest approach is no less fallible. Consider again the case in which each of ten mapmakers surveys precisely the same ten subregions – in this case, the calculation goes through just as in the previous paragraph, delivering $P(M) = 1$, but this is hardly the correct probability given the existing evidence. One can imagine many other scenarios in which the mapped subregions of the various mapmakers partially overlap, producing a weighted average that effectively accords too much weight to any sub-region which is mapped by more than one mapmaker. Thus it appears that if the modest approach is to work for cases like these, we must require that agents garner bodies of evidence which are disjoint, i.e. do not overlap – suggesting that this approach will work best when the accumulation of evidence is organized into a cooperative effort in which agents do not duplicate each other's evidence.[5]

But there is another snag. Heretofore we have been assuming that M is the true map – let us now suppose that M is false. In this case it is perfectly possible that even our modest approach to weighted averaging will fail to deliver a rational update even if the various bodies of evidence are disjoint. Suppose that just one of the 100 subregions of M is false – the other 99 are true. Now providing that none of the mapmakers actually surveys the false subregion, each mapmaker will post probability 1 for the subregion surveyed – and the weighted averaging method will proceed unhindered as above. But if any one mapmaker surveys the false subregion, she will post 0 for this probability. But in this event the technique of weighted averaging is undermined, for the only rational update for the probability that M is a true map is 0 (not the weighted averaging of the various p_i!) Here is yet another limitation on the modest approach – it fails in cases in which any agent posts probability 0 based on evidence that conclusively refutes the theory. The solution to the problem, of course, is simply to allow that

---

[5] The example as modified into the modest approach has some structural resemblance to an example of weight assignments treated briefly by Lehrer and Wagner (1981: 138–139): 'Suppose that n individuals are given a collection of N objects, each bearing a label from the set {1,2,. . .,k}, and must determine the fraction p_j of objects bearing the label j for each j = 1, 2,. . ., k. Suppose further that the collection is partitioned into n disjoint subsets, and that individual i examines a subset with N_i objects and reports, for each j = 1,2,. . .,k, the fraction p_{ij} of objects in that set which bear the label j . . . [T]he rational sequence of weights $w_1, w_2, \ldots, w_n$ is immediately apparent. Individual i should receive a weight proportional to the size of the set which he examines, so that $w_i = N_i/N$. For assuming that individuals count correctly, it is each to check that $p_j = w_1 p_{1j} + w_2 p_{2j} + \ldots + w_n p_{nj}$.' While L&W offer this as one example of how weights might actually be assigned, they do not make mention of the fact that what is going on in this example differs sharply from the use of weighted averaging to calculate the probability of some theory, which is of course the focus of their book; rather, what is calculated are various relative frequencies. (One might imagine that the agents are calculating the probability that a randomly selected object has label j.) Neither do L&W note that the technique of weighted averaging in this example differs from the technique applied in the rest of their book, insofar as the latter technique has various agents posting probabilities that a single theory T is true. Here, each agent merely reports information specifically about some proper segment of T's domain (and this information is combined to compute relative frequencies for the whole domain) with no reference to the probability of any particular theory. This is the approach I have deemed the modest approach.

P(M) goes to o whenever any agent declares that she has evidence that straightforwardly falsifies the theory.

So far we have considered a case in which agents independently seek bodies of evidence that they do not share with one another – I noted above that one might regard this as unlike the actual practice of scientists insofar as evidence sharing is a common practice. One aim of the publication or presentation of a scientific paper, of course, is to share evidence one has acquired with one's epistemic community, and evidence can be shared in more informal ways as well. One possibility is that, for some epistemic community considering the probability of some theory T, there is complete evidence sharing – so all members post probabilities for T based on the same all-inclusive body of evidence. This could result in members posting differing probabilities for T, of course, on the assumption that their various background theoretical beliefs differ. But currently we are assuming a shared theoretical background – in which case we would (barring logical errors in probability computation) expect all agents to post the same probability for T. In this case, one could use the original weighted averaging technique discussed above, assigning an equal weight to each member – for as noted there, this technique works when evidence is completely shared. Another possibility, one that I think more closely resembles actual epistemic communities, is to assume that there is a common body of evidence C that is widely known throughout the epistemic community – this could be evidence that has been presented in the past in prominent places and with which any competent member of the community should be acquainted. Individual members take C into account in considering the probability of T – but also take into account private bodies of evidence which they have acquired in their own research but which are not publicly known. This could be hard data they have acquired which has not been published – or perhaps evidence whose effect is to produce an intuition about the probability of T that guides the member's judgment yet remains hard for the member to articulate (i.e. 'hunch-producing' evidence). The pervasiveness of epistemic pluralism substantiated in Chapter 2 suggests that pluralism is an important phenomenon in theory evaluation in part because complete evidence sharing is sometimes difficult for epistemic communities to attain (a point for which Speelman (1998) shows there is considerable empirical evidence, cf. (2.4.3)).

To transform the mapmaker example into this sort of case, let us assume that there is a set of subregions that have been antecedently mapped and the results of this mapping been dispersed to all our X-many mapmakers – call this common body of evidence C (which clearly accords with M, which we assume once again to be the true map). Besides the possession of C, however, each member i possesses a privately held body of evidence R_i which amounts to one or more additional mapped subregions. If it should happen that R_i is the same subregion(s) for each member, then the original weighted averaging will work if we set weight w_i equal to the number of subregions regions mapped by i (the number of subregions in C and R_i together) divided by the total number of regions mapped by all mapmakers (X times the sum of subregions in C and R_i) – for this turns out to be just another example of complete evidence sharing. This method will not work, of course, if private evidence is disjointly held, for the same reasons it did not work in the case without evidence sharing: it will not allow the epistemic significance of the various disjointed bodies of evidence to be combined. To handle this sort of case we need a version of the modest approach. My proposal is as follows: because it is assumed that all members are familiar with C, and because all theoretical beliefs are shared, an agent who knows no more than C is accorded 0 weight. Such an agent essentially contributes nothing to the project of computing a communal probability for T. Agents are thus weighted proportionately to the amount of their privately held evidence. Where agent i has privately mapped n-many regions, $w_i = n/N$, where N is the total number of acts of private subregion mapping performed by all X-many members. We assume now that all privately held evidence is disjoint – there is no overlap in the various R_i. We use again the modest approach: each agent either posts 0 or 1 depending on whether their disjointly surveyed region of the city corresponds to M – as we are assuming M is the true map, each agent posts 1. The simplest way to compute a rational probability for M is to assume that, along with the X-members of the community, there is a phantom X + 1th member who possesses C, and has a weight equal to the number of subregions in C divided by the number of acts of private subregion mappings (on this reconstrual, the X + 1's mapping of C counts as a private mapping – thus the weights of the other agents must be computed on this assumption). This transforms the scenario which allows for a pool of publicly known evidence to be handled by the same modest approach that worked for a scenario in which evidence is held entirely disjointly. This suggests that actual scientists familiar with C and with the judgments of scientists whose private expertise has been weighted will post probabilities in a similar way: they take into account the evidential impact of C, given its size, and separately take into account the judgments of scientists that their privately held bodies of evidence support M (including their own, if they are among the evidence collectors) – the modest approach allows these disparate forms of evidence to be combined.

What if the various bodies of privately held evidence themselves partially overlap? This possibility could occur, e.g., if just two of the weighted agents held exactly the same privately held evidence – in which case it would make sense to exclude one of them from the weighting process. But this sort of practice is reasonable in actual cases where there is reason to think that private evidence is completely duplicated – say for example if the two are collaborators, or if one of the two is the student of the other, and can be expected to know just what the teacher taught him. In other cases partial overlap of private evidence can occur simply by chance – say because each happened to pursue a similar course of evidence-gathering (e.g. performed similar experiments). Here I admit that the technique of weighted averaging I propose may go wrong, for reasons noted, but of course it may go wrong for similar reasons in the actual computation of communal probabilities by actual agents. To the extent that evaluators can gauge the degree of evidential overlap, evaluators who are attempting to assign reasonable weights may be able to re-calculate such weights so as to base them on disjointly held evidence.

In this section we have considered a case in which the members of a community of knowledge seekers begin with a common body of background belief Kt and independently explore a large and diffuse body of evidence – either with or without evidence sharing. We conclude at this point that the technique of weighted averaging, refined into the modest approach, may be of some use in cases like these, but only under some fairly stringent conditions. These include (1) the amount and epistemic impact of disjointly held bodies of evidence is roughly known (so that various w_i reflect the amount of disjointly held bodies of evidence), and (2) that all the garnered evidence effectively supports, rather than falsifies, T (this will occur in any scenario in which a community tests hypothesis h by testing the observable consequences of h and each of them is confirmed). Where condition (2) is not met, weighted averaging will fail, but the necessary judgment about T's probability is easy to make: in this case T has probability 0.[6]

---

[6] There is of course another sort of case: one in which evidence is produced by an agent that neither confirms nor refutes the theory at issue but merely disconfirms the theory to some degree. I leave aside this complication here, as in situations like our mapmaking scenario it is presumably easy to see whether any accumulated evidence accords with M or not – if it does, M is supported, if it does not, M is refuted (i.e. there is no likely occurrence of evidence that merely 'disconfirms' the theory to some

The fundamental intuition about weighting with which we began this section was that weights should reflect the amount of evidence garnered by the weighted agents. But as matters have turned out, this is not quite the correct gloss on weighting in the mapmaker scenario. Following the modest approach, and absent complete evidence sharing, weights should reflect not merely the amount of garnered evidence, but the extent to which the body of evidence extends the totality of evidence garnered by the total community. Agents are not weighted thus on the basis of how much evidence they possess, but on the basis of how much evidence they have garnered that has not been garnered by others. This should not suggest that scientific communities do not assign credibility to agents on the basis of the agents' possession of publicly available evidence – to the contrary, I would argue that acquaintance with publicly available evidence is a necessary condition for membership in a scientific community. But the weighting process is applied only to individuals who have been admitted and purports to measure what such individuals can add to what is already known about the evidence that bears on theories. Let us now turn to a quite different scenario.

6.4 SCENARIO TWO: THE PROBABILISTIC JURY

Let us consider a person charged with tax evasion – he undergoes a trial by X-many jurors. During the trial a carefully circumscribed body of evidence is presented to the jury consisting of documents, letters, testimony about the defendant's background, etc. The jurors all listen carefully to all the evidence (which becomes Ke_i for each juror i). We assume, however, that jurors may bring different background beliefs to the trial – these will be the sort of background beliefs that jurors draw on in assessing the probable guilt of an accused person: e.g., beliefs about the relative frequency of the crime at stake in the relevant community, about the probable soundness of various arguments provided by the defense and prosecuting attorneys, about the relation between a witness's behavior on the stand and the trustworthiness of that witness, etc. I will refer to each juror i's background beliefs as Kt_i (these are, in this context, their 'theoretical' beliefs). Now let us suppose that the jurors are not required simply to declare the defendant

degree). Theory testing tends to have this feature when the theory at stake is a description of some epistemically accessible domain – and the evidence likewise may be easily surveyed and requires no interpretative skill to understand. I discuss the difference between theories with and without this sort of feature – and the different ways in which their communal probabilities are computed – in more detail below.

either guilty or not guilty – rather, after the jurors collectively deliberate, each juror is required only to post a probability for the statement (G) The defendant is guilty. Our problem is whether and how averaging might be used to provide a rational update for P(G) given the X-many posted probabilities.

In the mapmaking scenario we assumed on the part of our various agents an identical starting point (i.e., their theoretical background beliefs are the same) and a subsequent independent search that provided various pools of evidence. In this case, we assume potentially different theoretical background beliefs but only a single, shared body of evidence. In so far as it is at the level of the background theoretical beliefs Kt_i that our jurors differ, in weighting the jurors we are somehow weighting their respective Kt_i. But how would such weights be fixed? In the mapmaking scenario, an agent's weight reflected the amount of empirical evidence that he garnered that was not held by other agents. Agents in that scenario thus could be thought of as collaborators whose various evidence-gathering activities are collated by the weighted averaging technique. Thus the weighting technique was adjusted so as to produce a communal probability that was based on the total available evidence for the relevant theory. But something quite different is going on in the case of our probabilistic jury. Given the potential divergence between the agents' various background theoretical beliefs Kt_i, the relationship between the agents seems less collaborative and more competitive – they have potentially conflicting views about the subject at hand and the weighted averaging technique must somehow serve to adjudicate this conflict. The weighted averaging technique thus cannot be thought of as a device for collating evidence, nor does it seem right to think of weights as based upon disjoint evidence possession. How should it then be conceived?

Insofar as the agents on the jury differ at the level of their theoretical beliefs, weights should somehow measure the merit of those theoretical beliefs. Now of course we have developed in considerable detail an account of predictivism that serves to measure the probable truth (and more recently, the probable sufficiency) of an agent's background beliefs – thus it seems we should be well positioned to compute weights for agents, provided their record as endorsers is available to us. However things are more complicated than this. It is important to remember that the account of predictivism developed in Chapter 3 provides us with a rather limited basis for assessing an agent's background beliefs K – for, as noted above, it assesses only whatever background beliefs bore on some particular theory T. As noted above, when we turn to consider the epistemic significance of some posted probability for some new theory T', we should avoid assuming that the background beliefs K' that bear on T' are entirely the same as those K that bore on T. Insofar as T and T' are theories in the same scientific field, it may well be that there is partial overlap between K and K'. Absent even partial overlap, it may be that the content of K' lies in epistemic proximity to that of K, such that an agent who possesses K is more likely to possess K' than someone who lacks K.[7]

Consequently it seems reasonable that a sequence of novel successes across a sequence of theories which all fall within the same field F functions as evidence of the overall credibility of that agent's background beliefs with respect to F – and that this sort of overall credibility is what matters when we consider the merit of the agent's posted probability for some new theory within F. But nothing in the previous chapters provides us with any obvious procedure for computing this sort of overall credibility.

Another limitation of our account of predictivism is that it is focused on a particular kind of agent, the endorser (either a predictor or accommodator). But these are not the only kind of agents who are important to consider in formulating a general account of epistemic pluralism – for we are concerned with the credibility of agents who have posted non-endorsement probabilities for theories as well. An agent who posts a skeptical or non-committal probability for some theory should be accorded greater weight than an endorser of the theory if it is later repudiated.[8] Clearly, we need more resources for agent evaluation at this point than our account of predictivism has provided us with.

What is important about agents in our current context is their disposition over a sequence of probability postings to post probabilities that are accurate. The accuracy of a probability is inversely proportional to its distance from the truth value, where I construe the truth value 'true' to be 1 and the truth value 'false' to be 0. Of two agents who have posted probabilities for a sequence of theories, all of whose truth values have been subsequently determined beyond reasonable doubt, the more credible agent is the one who has on average posted more accurate probabilities. The more credible agent should thus be weighted more heavily in the computation of a communal probability.

My proposal is that in this scenario we think of agents as probability posting devices whose degree of accuracy relative to one another can be calibrated. Given that each agent i posts a probability for G (given the

---

[7] Compare Barnes (1996a: 405–407).
[8] An agent who posts a probability so low as to be considered a repudiator of the theory is really an endorser, for he endorses the negation of T.

shared body of evidence presented in the trial) on the basis of Kt$_i$ (together with the shared evidence) such a calibration for agent i effectively functions as a calibration of the degree of credibility of Kt$_i$ relative to the other agents. Provided that such a calibration is properly performed, the rationality of weighted averaging can be demonstrated. Weighted averaging thus construed does not serve to collate evidence (conceived as observational knowledge that bears directly on the theory at stake) but to collate information about the relative accuracy of various probability posters who have posted particular probabilities for the theory at stake.

At this point I propose that a suggestion of Lehrer and Wagner's might be developed to suit our current purposes. Toward the end of their 1981 book they quickly sketch several suggestions for how weights might be calculated – see Footnote. 5 for discussion of one of these. Here is another:

Suppose that n individuals are attempting with unbiased devices of differing accuracy to measure a quantity u. Assume further that their estimates a_1, a_2, ..., a_n of this number may be regarded as realizations of a sequence of independent random variables X_1, X_2, ..., X_n with variances σ_i. Variances are often used as 'performance' measures in estimation problems, reliability varying inversely with variance magnitude. If this measure of reliability is adopted by the group, the group should adopt as their consensual estimate of u the number $w_1a_1 + \ldots + w_na_n$ where the weights w_i are chosen to minimize the variance σ^2 of the random variable $w_1X_1 + \ldots + w_nX_n$. It is easy to check that $\sigma^2 = w_1^2\sigma_1^2 + \ldots + w_n^2\sigma_n^2$, and a little partial differentiation then shows that σ^2 is minimized when $w_i = (1/\sigma_i^2)/(1/\sigma_1^2 + \ldots + 1/\sigma_n^2)$. In decision making contexts like the one under discussion, one might initially have been inclined to endorse the estimate of the individual with the smallest variance as the rational group estimate. Yet such a policy is demonstrably inferior (with respect to variance minimization) to the use of the above weighted average, indicating the wisdom of collective deliberation on a single matter, even when some individuals are more expert at this matter than others. (1981: 139)

(Readers interested in a proof of the claim that these weights indeed serve to minimize variance are directed to Hoel (1971: 128–129).) Lehrer and Wagner have indeed cited an example in which weighted averaging may be rationally deployed, though they do not note the fact that this example seems not to be one in which what is estimated is a probability for a theory – rather, what is estimated is the value for some magnitude. How might this proposal be developed so as to apply to probability estimation? We might deploy this apparatus with the proviso that the measured magnitude is the 'correct epistemic probability' of some theory (like our theory G) given existing evidence, and where such correct probabilities exist (and become subsequently known) this may be a viable strategy. But the assumption that correct probabilities exist is a quite troubled one in many circumstances, including circumstances in which experts hold differing background theoretical beliefs and thus differ about probability assignments (and such circumstances are precisely where we might hope weighted averaging to be most useful).[9] Thus I propose that the estimated magnitude be regarded as the truth value of the theory T whose probability is being estimated, where the truth value is 1 if T is true and 0 if T is false. The posting of probabilities thus becomes a kind of truth value 'measurement' of sorts: the closer a posted probability is to the truth value of T, the better the measurement.

Let us now return to our jury story and enrich it with some additional assumptions. Let us suppose that each of our X-many jurors have in fact previously served on juries for many similar cases. Suppose there is a professional jury system in which jurors specialize in adjudicating a particular sort of trial – each of our X jurors, let us assume, has served as a juror for many similar tax evasion trials (including trials of a similar degree of 'difficulty' with respect to how much deliberational skill is required to deliver a wise probability). For some significant subset of these trials, moreover, the eventual guilt or innocence of the defendant was determined beyond doubt (in each case, after the jury had posted its probabilities). These assumptions allow us to apply the minimal variance weighting technique to our probabilistic jury. Each juror J, we now suppose, has a variance σ_j which is inversely proportional to that juror's reliability as a poster of probabilities. We thus calculate minimal variance weights in the way sketched above – and use weighted averaging to determine an updated probability that the defendant is guilty. We are assured thus that the variance of the weighted average will be less in the long run than that of any particular juror, and the rationality of weighted averaging for probability updating is, in this precise sense, assured.

Whereas the weighting technique in the mapmaking scenario measured weights in terms of disjoint evidence possession, this scenario imagines that

[9] To expand: a correct epistemic probability of T will exist when there are codified and uncontroversial procedures for calculating the probability of T together with a well defined and uncontroversial body of evidence relevant to T. Consider, for example, the probability that an agent will win a fair lottery with n tickets sold having purchased one ticket – the correct epistemic probability for this claim is 1/n. The point here is that such correct probabilities will of course often fail to exist, either because there is no codified procedure for probability estimation or because the background beliefs with respect to which probabilities are computed differ among competent agents. The non-existence of 'correct epistemic probabilities' is a familiar point to Bayesians (and critics of Bayesianism): it is just the familiar point that there is no general procedure for fixing prior probabilities.

the evidence (i.e. the information presented in the trial) is shared between all participants – the weighting technique reflects our estimate of how likely it is that an agent will deliberate upon that evidence so as to produce a probability close to the truth value of G (the proposition that asserts the defendant is guilty). Because probabilities are posted, for a given body of evidence, on the basis of an agent's background beliefs, I interpret minimal variance weights to be a measure of the sufficiency of an agent's overall theoretical background beliefs relative to other agents.

The proposed method of weight computation, moreover, incorporates our intuition that weights are partly measured by the novel successes of agents – for the track records of agents with low variance can include cases in which agents posted relatively high (viz., endorsement level) probabilities for theories which were later counted as true due to their novel successes. It is important to emphasize that this essentially amounts to a new version of predictivism – but it does not compete with the account developed in previous chapters. In the previous account we begin with a prior probability for the background beliefs K which bear on the probability of T and update the probability of both T and K accordingly. In this new account we extrapolate from past predictive success (along with past probabilistic accuracy in general) to the greater probability of accurate probabilities in some new case. One way of understanding the relation between the two predictivisms is to think of the new predictivism as supplying evidence that is relevant to the establishment of the prior probability of background belief required in the old predictivism. For example, the various prior probabilities that the agents' background beliefs are sufficient (should they subsequently endorse some set of competing theories) should be ordered in accordance with the ordering of their weights as determined by the method presented here.

6.5 APPLICATIONS FOR THE TWO SCENARIOS

The mapmaker scenario amounts to one in which a group of individuals individually compile bodies of evidential information which the weighted averaging technique seeks to combine. One sort of case of knowledge pursuit that has this structure is the sort performed by the sorts of research teams described in Hardwig (2001), in which individual team members contribute evidential information without ever coming to possess (in the sense of 'grasp') the evidence procured by other team members. The individual pieces of evidence are assembled to produce a body of evidence that establishes some theoretical claim. But more mundane applications are also imaginable: imagine an historian who presents a long and detailed narrative on, say, the history of Australia. The community of historians seeks to assess the probability that the narrative is correct (or at least approximately correct). We may well suppose that the narrative will be read by various Australian historians with disjoint areas of expertise, their expertise could be divided along temporal, regional, or other lines.[10] Each reader is weighted with respect to the relative quantity of expert knowledge he or she (disjointly) possesses vis à vis others, and then each historian declares that the account is accurate or not with respect to his area – thus weighted averaging (specifically the modest approach) could proceed more or less as in the case of our mapmakers, with the same conditions on its application.

The probabilistic jury corresponds to a case in which agents begin with diverse theoretical beliefs but then reflect on a unique body of evidence. Thus this example is structurally similar to cases in which a well-defined body of evidence is reflected upon by a group of agents with somewhat different backgrounds, all presumably competent by some minimum standard. Differences in assigned weights, as I understand them, reflect differences in assessed theoretical background belief sufficiency rather than differences in the evidence possessed. Examples would include scientific communities consisting of agents who each consider how likely a particular theory is based on a shared body of evidence, but also any group of experts who must collectively post a probability based on shared evidence (engineers declaring the probability that a certain project can be accomplished under a certain price, physicians declaring the probability that a patient will recover from a certain illness under a certain therapy, etc.) There can be little question that such agents are evaluated by their peers with respect to their relative epistemic authority, and this proposal shows how such evaluations may be rationally deployed in communal probability updating.

---

[10] In so far as the mapmaking scenario was one in which we imagined the various mapmakers start with the same background theoretical beliefs, we must tell this story as follows: the various readers begin their study of Australia with the same background theoretical beliefs (as they pertained to Australia's history), and thereafter acquire disjoint bodies of historical information prior to assessing the narrative of our example. It is not essential to the example that their historical knowledge literally be disjoint, of course – but only that each historian evaluates the narrative from the standpoint of his or her particular area of expertise, which we assume is different from the other historians. Thus the evaluation falls in line with the modest approach – and of course the existence of a common shared pool of evidence can be accommodated in the way suggested in Section 2 above.

6.6 MIXED WEIGHTS

We have explored two strategies for fixing weights, and now must confront the following question: given some actual scenario involving multiple agents who have posted probabilities for some theory, how would an evaluator choose which weighting method to use? The answer depends, apparently, on what sort of information is available in the actual scenario. If we possess information about the extent to which the various agents are (disjointly) in possession of the relevant evidence, we will use the evidence-based method used in the mapmaking scenario. If we possess information about agents' track records in posting probabilities for similar theories, we will presumably choose to use the track-record based method. Lacking either kind of evidence, we should probably refrain from assigning any weights at all.

But suppose we have both sorts of information available. How would these two sorts of information be combined to compute a set of 'mixed weights' which would incorporate both pools of information? It might be tempting to simply assign two weights to each agent, one based on each weighting technique, and then declare the average of the weights to be the agent's actual weight (a community's set of weights thus computed will of course sum to one). This procedure, however, is too simple. It would effectively assume that in this actual scenario, the types of information about the agent's competence are of equal importance. But this assumption may well be false, as we will now see.

There are cases in which what matters primarily (with respect to the credibility of an agent's probability) is not how much evidence the agent possesses that is not possessed by others but rather how skillfully he can assess a given body of evidence (the kind of skill that the minimal variance weights purport to measure in the jury scenario). There are other cases in which this kind of skill is less important than the possession of evidence that is not possessed by others. The primary differences between two sorts of cases concern (1) how much depth of theoretical knowledge is required to competently assess the significance of a given body of evidence and (2) how likely it is that the procurement of new evidence will improve the probability estimation process. Cases in which theoretical knowledge is more important than the possession of evidence not possessed by others include cases in which it is judged that additional evidence is unlikely to affect the probability estimation process very much. Consider a group of pharmacologists who are interested in the toxicity of a particular substance on human beings – they have surveyed the results of extensive and thorough toxicity testing of the substance on a variety of laboratory animals. Now a maverick pharmacologist arrives and declares that he alone is in possession of pertinent evidence that he has acquired by testing the substance on various animals that are not standardly used in laboratory testing – he has tested the substance on ostriches, panda bears, and moles. Even if he truly claims to be the sole possessor of a wealth of such information, he may not receive a weight proportionate to the sheer quantity of this evidence. Such evidence may well be judged to be, while not irrelevant, more or less superfluous given the totality of evidence on standard laboratory animals that already is publicly accessible. What matters in this case is the ability to competently assess the publicly known evidence, and pharmacologists should be weighted primarily (though perhaps not exclusively) on this basis.

Cases in which the possession of evidence not possessed by others is more important than theoretical knowledge include cases with two features: first, little theoretical knowledge is required for the competent assessment of evidence, and second, the evidence uniquely possessed by the agent is believed to be of considerable importance to the estimation of the probability. The mapmaking scenario is clearly intended to be a case of this type, but it is not hard to imagine others. Imagine a group of anthropologists who are interested in the culture of a particular native community – the anthropologists have differing degrees of theoretical knowledge. One of these anthropologists, however, has had extensive contact with the members of this community – she is fluent in their language and has lived among them for years. She has, for example, been extensively exposed to the folklore of the community. Her views about the folklore of that community should thus be heavily weighted without much reference to her theoretical knowledge qua anthropologist. This will be true, at least, if the content of a culture's folklore is something that can be thoroughly known without bringing to bear the sort of theoretical knowledge that might be valued more highly in some other context, and because such first hand acquaintance with folklore is indisputably more likely to improve the estimated probability of some account of that folklore.

We thus require a method of calculating mixed weights which is sensitive to local distributions of emphasis between evidence possession and theoretical knowledge. Such a method will of course only be applicable in cases in which both sorts of information are available regarding each agent in the community. In such cases I propose that each agent i receive two weights, w_{ie} and w_{ik}, which are determined according to the evidence-based and theory-based methods respectively. Agent i's mixed weight m_i will thus be a weighted average of these two:

$$m_i = aw_{ie} + (1 - a)w_{ik}. \qquad (4)$$

Coefficient a reflects the amount of emphasis accorded to disjoint evidence possession relative to theoretical knowledge.

6.7 WEIGHTING GONE WRONG

In Chapter 1 I identified the paradox of predictivism as consisting of a pair of inconsistent but deeply felt intuitions: on the one hand it seems as though the act of prediction by a human agent could not carry epistemic significance, on the other hand it seems as though it must. I have argued that one resolution of this paradox can be found in epistemic pluralism – the position that human judgments can and often do carry epistemic significance in many different types of scientific communities. For predictions are, on my view, judgments of a certain kind. Philosophers who deny epistemic pluralism will be led to deny predictivism (at least in its tempered form). In Chapter 2 I defended the claim that pluralist theory evaluation is both pervasive and perfectly reasonable – and in this chapter I have attempted to demonstrate more rigorously how pluralistic theory evaluation can proceed in a rational manner.

But for all that has been said, some readers will perhaps still object to the legitimacy of pluralism in genuinely scientific contexts – and such readers will find something objectionable in the technique of weighted averaging that I develop in this chapter. On their behalf I propose to consider two basic objections to the weighted averaging project, which are in fact just objections to the legitimacy of epistemic pluralism. These are (1) it is fundamentally unscientific to estimate probabilities by weighted averaging because such a method will promote an unhealthy elitism in which the views of prestigious scientists are accepted merely on the basis of prestige-indicators like institutional affiliation, and (2) the weighted averaging technique misunderstands the way in which attributions of prestige function in scientific deliberations. We deal with each in turn.

On objection (1): one might imagine that one way in which weights are distributed is on the basis of 'prestige,' where this quality is reflected by factors like their institutional affiliation or educational pedigree. But to privilege scientists opinions (including their probabilities) on the basis of factors like these is blatantly unscientific – it will lead to an uncritical acceptance of certain views merely because they are propounded by 'prestigious' people. To illustrate: Peters and Ceci (1982) present evidence that scientific papers in one discipline were accepted on the basis of the institutional affiliation of the authors rather than the intrinsic merit of the papers. Helen Longino (1990: 68) comments that 'Presumably the reviewers . . . assume that someone would not get a job at X institution if that person were not a top-notch investigator, and so his/her experiments must be well done and the reasoning correct.' At first blush, such a case might seem an excellent datum with which to argue against the use of weighted averaging in general. 'See what happens,' a critic might insist, 'when assessments of scientific claims are grounded not on a careful examination of the evidence on which they are based but on an assessment of the so-called credibility of the investigators.'

However, such a criticism can be turned around. The cause of the problem we are describing, in fact, is not that the credibility of various agents with differing views would be taken account in assessing the relevant scientific claims. It is, rather, that too few agents are weighted in the assessment – only 'prestigious' authors are weighted – and their weight is assessed all too simply on the grounds of prestige-indicators like institutional affiliation (which is not to deny that this could be one source of information about credibility, albeit a fallible one). The solution to defective weighting procedures is not to abandon the use of weighting in general, I would argue, but to improve the calculation and distribution of weights. The remedy for such elitism is a more democratic weighting procedure that spreads cognitive authority across a greater variety of agents, recognizing the often legitimate claims to authority made by scientists at institutions not generally recognized as prestigious.[11]

On criticism (2): perhaps it is the case that although an agent's credibility, as assessed by others, clearly plays some role in scientific dialogue about

---

[11] It should be noted in passing that one of the central theses of Longino's book *Science as Social Knowledge* squares neatly with the argument presented here for epistemic pluralism. Longino argues that the objectivity of science consists in large part in the fact that scientific dialogue occurs between a plurality of agents with differing perspectives. She writes that 'only if the products of inquiry are understood to be formed by the kind of critical discussion that is possible among a plurality of individuals about a commonly accessible phenomenon, can we see how they count as knowledge rather than opinion' (1990: 74). She concludes that 'Scientific knowledge is, therefore, social knowledge. It is produced by processes that are intrinsically social' (ibid.: 75). Although Longino does not discuss the use of weighted averaging itself (she focuses primarily on interpersonal dialogue about method) her analysis is applicable to it in general. Her argument that the scientific community requires the incorporation of a variety of 'voices' is a call for a particular sort of weighting procedure. Longino's community-based conception of inquiry stands in sharp contrast to the conception of 'epistemic individualism' which Hardwig claimed was at the heart of traditional epistemology (cf. 2.1). A related thesis of Longino's – that scientific objectivity requires that scientific community incorporate the voices of women and minorities – can also clearly be understood as a call for a more democratic weighting procedure (1990: 78f)

claims that agent makes, that role is not accurately portrayed in terms of weighted averaging. Rather, an agent Y's assessment of another agent X's credibility serves to determine how much of a hearing Y accords to X's views on the matter at hand. If Y assigns a high weight to X, this means that Y will pay careful attention to X's views in determining Y's own views – conversely, if Y assigns X low weight, Y will not waste too much time in pondering the evidence and arguments presented by X. This sort of weighting is ultimately consistent with epistemic individualism, for while Y may rely on X to provide Y with evidence and arguments Y would not have considered on her own, ultimately Y's probability may reflect only Y's assessment of all relevant evidence (rather than a process of weighted averaging).

No doubt this sort of 'weighting' does transpire – but in my view it cannot be the whole story in general. For, as argued previously, ultimately rational agents must concede that there are some contexts in which, try as they might, they will never absorb all the knowledge, arguments and expertise their colleagues possess, even on subjects on which they are themselves experts. Such rational agents include even expert evaluations (such as the interdisciplinary, imperfect, humble and reflective evaluators of Chapter 2). At such points there is no choice but to concede that they must assign a non-zero weight to their colleagues' probabilities even when they disagree with the probabilities their colleagues post. One could only do otherwise by effectively assigning zero weight to colleagues despite the impressive credentials those colleagues may possess. While some scientists may choose this path, I would argue that the wisest ones do not.

Philosophers who take seriously the hypothesis that scientific discourse is and ought to be pluralistic should not ignore the technique of weighted averaging in probability estimation. Though the technique is no panacea, it provides a rigorous tool for representing one way in which differing perspectives in inquiry clash and mesh with one another. Pluralism emerges as a rational form of theory evaluation, and this lends credence to the position that predictivism – in its tempered form – is often times a rational procedure for theory assessment.[12]

---

[12] Some of the results of this chapter were presented in Barnes (1998). Paul Meehl's (1999) response to this paper argues that the general problem of fixing weights is not a particularly serious one. He argues that "if scientists' judgments tend to be positively correlated, the difference between two randomly weighted composites shrinks as the number of judges rises. Since, for reasons such as representative coverage, minimizing bias, and avoiding elitism, we would rarely employ small numbers of judges (e.g. less than 10), the difference between two weighting systems becomes negligible." (1999: 281) Meehl makes suggestions for quantifying verisimilitude, identifying 'types' of scientists or theories, inferring latent factors, and estimating reliability of pooled judgments.

CHAPTER 7

Postlude on old evidence

7.1 INTRODUCTION

In 1927 Bertrand Russell delivered a characteristically subversive lecture entitled "Why I am not a Christian." In 1980 Clark Glymour published an essay with the clever title "Why I am not a Bayesian." Glymour's critique of Bayesianism resembled Russell's critique of Christianity in arguing that it was a view that was far more popular than it should be, given its merits. Among the various problems suffered by Bayesianism was one Glymour called the problem of old evidence.

To say that evidence e confirms hypothesis h on a Bayesian analysis is, intuitively, to say that $p(h/e) > p(h)$. However, when e is antecedently justified (i.e. when e is 'old evidence') and $p(e) = 1$ it follows that $p(h/e) = p(h)$ (assuming e is a logical consequence of h).[1] Thus it appears that old evidence cannot confirm a hypothesis and – glory be – predictivism is vindicated! Not only predictivism, but the null support thesis – the radical Popperian view that old evidence is powerless to confirm a theory – seems to be straightforwardly established. But a moment's reflection suffices to realize that we have proved too much – for it is incontestable that old evidence can offer confirmation of a theory, as we saw, for example, in the case of Mendeleev's periodic law. But then it appears that, from a Bayesian point of view, the inequality $p(h/e) > p(h)$ does not capture what it means to say that e confirms h for old evidence e. More specifically, the Bayesian inequality fails (1) to entail that old evidence is evidence (the qualitative problem) and (2) to establish to what degree old evidence is evidence (the quantitative problem). The Bayesian problem of old evidence is to explain what conditions do properly capture the qualitative and quantitative aspects of confirmation for such cases.

---

[1] The point follows immediately from Bayes' theorem: $p(h/e) = p(h)p(e/h)/p(e)$.

Aside from the qualitative/quantitative distinction, the problem of old evidence has been further analyzed into the synchronic and diachronic versions of the problem.[2] The synchronic problem concerns the support provided by evidence E for T at some single point in time t after E has come to be known and T has been formulated, and their logical relationship established. The diachronic problem concerns the effect that taking E into account had on an agent's probability for T at that particular moment.[3] In this chapter we will be concerned with just the synchronic problem of old evidence.

There are several reasons why the old evidence problem cannot be ignored in our present context. One of these is that, as noted, insofar as $p(h/e) = p(h)$ for known e, we seem to have a simple Bayesian argument for a radical form of predictivism, but one that is intuitively unsound. In so far as it is our aim to understand predictivism in general we should be concerned to disqualify bad arguments for our thesis – thus we require a solution to the old evidence problem that explains precisely how the putative argument for predictivism goes wrong. But there is a more important reason for taking up this problem now, and this specifically concerns the quantitative problem of old evidence. As we will see, there is a significant body of opinion that holds that the quantitative problem has no solution, and that there is no fact of the matter to what degree some known e confirms a hypothesis h for some agent. If this were true, the thesis of predictivism would be threatened in so far as this thesis is designed to apply to cases in which evidence is known, and given that this thesis makes a quantitative comparison between the degrees of confirmation provided by predicted vs. accommodated evidence. If predictivism is true, facts about degrees of confirmation must exist. Also, readers with excellent memories may remember that in (3.4) I claimed it was possible to 'purge' a body of background belief of its commitment to an empirical claim – and the solutions to the old evidence problem developed in this chapter enable one to do this. In what follows I develop two solutions to the quantitative problem of old evidence (one of which will turn out to be superior to the other). I also consider two solutions proposed by others and compare

---

[2] I follow Christensen (1999) here in my choice of terms – the distinction between the diachronic and synchronic problems of old evidence is identical to Garber's (1983) distinction between the historical and ahistorical problems of old evidence.

[3] One important approach to the diachronic problem has been to show that known E can provide an actual boost in the probability of T at some point because the logical relationship between T and E is established just at that point – thus the agent 'conditionalizes on logical discovery'. See Garber (1983), Niiniluoto (1983), Jeffrey (1991), (1995) and Wagner (1997), (1999).

them to my own. The upshot however is that the quantitative problem does admit of solution and thus does not threaten our predictivist project. Neither does the problem of old evidence itself point to a sound argument for predictivism for reasons that will be explained.

One neat solution to the qualitative problem was proposed by van Fraassen 1988: 154; cf. also Kaplan 1996: 49–51. If we impose the reasonable restriction that proposition e, which we assume to be a contingent statement, not be assigned probability 1 (a policy I have followed throughout this book), then the status of e as evidence for h can be established by the inequality p(h/e) > p(h), and the qualitative problem of old evidence does not arise. But existing literature contains no adequate solution for the quantitative problem. To the contrary, as noted above, there is a near consensus among recent authors that the quantitative problem may be unsolvable. Below we will briefly review the current status of the discussion of the quantitative problem of old evidence.

Most existing attempts to solve the quantitative problem have been based on a particular approach to this problem I deem the 'e-difference approach.' The basic idea is as follows: we assume agent A possesses a stock of background belief K at time t, and that K includes e. On the basis of K A endorses a probability function p at time t defined on e and h (and all other propositions of A's language). To determine the degree to which e supports h for A at t, we should consider the (non-actual) scenario in which A does not believe e, but retains the rest of her belief – on this basis A selects a new probability function p∗. The degree to which e confirms h for A at t is equal to p(h) − p∗(h) – this degree is clearly based on the difference between p(e) (the probability assigned to e given that e is believed) and p∗(e) (the probability that is assigned to e on the assumption that e is not believed).[4] What essentially amounts to the e-difference solution was proposed by Glymour (1980: 87–91) only to be rejected by him (for reasons shortly to be discussed). A primary proponent of this sort of solution to the quantitative problem has been Howson 1984, 1985, 1991, 1996.[5]

As stated the e-difference solution is critically ambiguous. There are two feasible interpretations of what it means to assume that an agent who

---

[4] I use the method of subtracting a prior from a posterior as a confirmation measure in this chapter because it is simple, intuitively plausible, and because the question about the various virtues of competing measures does not seem to bear on any central issue here.

[5] Howson (1984) states the solution in terms of the e-difference approach roughly as presented here – his (1985) restates this solution along with a reply to Campbell and Vinci (1983). Howson (1991) develops the earlier proposal in a way that is defective for reasons noted in Maher (1996 ft. 10). Howson (1996) applies the e-difference solution to the example of Einstein in 1915 though he does not come to grips with the various issues I discuss in Section 3 below.

believes e does not in fact believe it – I deem these the 'counterfactual history interpretation' and the 'evidence deletion' interpretation. The counterfactual history interpretation of a non-belief assumption runs as follows: in assuming that an agent (who actually believes e) does not believe e, we are assuming that the agent's epistemic history was such that the agent did not come to believe e at any point, and his consequent epistemic history evolved as it would have had he in fact never come to believe e. But according to the evidence deletion interpretation of the non-justification assumption, in assuming that an agent who believes e does not believe e, we are not inquiring about 'what really would have happened' had the agent never come to believe e. Rather, we are assuming that the agent is hypothetically deprived of a piece of evidence (here, a belief) that she does in fact possess while retaining all other belief (irrespective of what history might have produced this state of affairs).

It should be noted that the evidence deletion approach is, despite its name, no less a counterfactual approach than the counterfactual history approach – the difference between the two approaches corresponds to different ways of interpreting counterfactual statements. The distinction between the counterfactual history interpretation and evidence deletion interpretation mirrors Lewis's (1979) distinction between two interpretations of counterfactual claims of the form "If A were true, then C." One possible interpretation (the one which supports 'backtracking arguments') argues that if A were true (at present), the past would be significantly different than it was. The other interpretation of counterfactual claims proposes that, in assuming that A is counterfactually true, the past is as similar to the actual past as possible, and is altered by the smallest possible miracle needed to make A true at present. The former interpretation corresponds to the counterfactual history interpretation, the latter to the evidence deletion interpretation (where A = Evidence e is not possessed and C = Hypothesis h has probability x). Significantly for our purposes, Lewis argues that the backtracking interpretation of counterfactual claims is most typically the wrong interpretation to impose on counterfactual claims. Curiously, as we will see, most critics of the e-difference solution have applied the backtracking interpretation – that is, they have conceived of the e-difference solution in terms of the counterfactual history interpretation. An argument of Eells is typical (1990: 208). He points out that in some cases the hypothesis h would not even have been formulated had evidence e not been believed – thus it makes no sense to inquire what probability A would have assigned to h had e not been believed (this type of criticism is strongly endorsed by Chihara (1987), van Fraassen (1988),

Earman (1992: 123) and Mayo (1996: 334)). In general, there may have been multiple effects that antecedent belief of e had on the evolution of inquiry relevant to h – and thus it may be extremely difficult to estimate what the general state of belief would be if e had not antecedently been believed. I agree that this objection is quite decisive when applied against the counterfactual history interpretation – but it does not apply to the evidence deletion interpretation. In what follows we will dispense with the counterfactual history interpretation and assume the evidence deletion interpretation of the e-difference approach to the quantitative problem.

But there is a threatening problem that does apply to the e-difference approach. Glymour's objection to this solution was that the mere supposition that A does not believe e is itself a vague supposition, for what does it mean for A to not believe e? In Bayesian terms, which prior probability should we imagine A assigns to e when we suppose that A does not believe e? Glymour's point is that there is no fact of the matter what this sub-justification prior should be – and thus no fact of the matter to what degree e confirms h for A from the standpoint of the e-difference solution. We refer to this below as 'the indeterminacy of $p*(e)$.' The indeterminacy of $p*(e)$ is stressed time and again in the literature (cf. Garber (1983: 103), Eells (1990: 207–208), and Earman (1992: 123)). Kaplan seems to speak for the majority in the Bayesian community when he concludes that there is no reason to think that there is anything in A's actual or counterfactual degrees of belief that enables one to measure how strongly e confirms h for A at time t (1996: 83). Mayo concurs (1996: 334), claiming that the proposal that some determinate value for $p*(e)$ could be fixed is 'silly.'

My position is that the apparent unsolvability of the quantitative problem (by way of the evidence deletion interpretation of the e-difference approach) arises in part from a particular widespread conviction about the nature of evidence within Bayesian circles: this is the idea that the only thing which qualifies as a 'piece of evidence' is a proposition, and such a piece of evidence is possessed by agent A insofar as it is believed by A. As long as we retain the view that evidence is exclusively propositional, we are destined to be befuddled by the quantitative problem as long as we are trying to solve that problem by way of the evidence deletion interpretation of the e-difference approach.

What might a piece of evidence be, if not a proposition? If by 'evidence for h' we simply mean whatever it is that supports the probability of h, then it is obvious that not all evidence can be propositional. If all evidence were propositional, then all justification would be inferential – but if all justification is inferential then justification is impossible in virtue of a well-known infinite regress argument. The argument, traceable at least to Aristotle's Posterior Analytics (Book I, ch. 2–3), is that if all justification were inferential, then the justification of any belief would require the antecedent justification of an infinite series of beliefs. The standard way out of the regress is to posit a non-propositional form of evidence, something like 'experience' or 'observation,' which serves as the terminal ground of empirical justification.

Careful readers of Bayesian literature have perhaps noted how little Bayesian authors typically have to say about observation – typically, observation is regarded by Bayesians as simply something that causes our probabilities to change. But observation is part and parcel of the paradigmatic Bayesian account of learning. According to standard Bayesian accounts, the paradigmatic case of learning occurs when an agent performs an observation (hereafter, an 'observational act') that serves to raise the probability of e (an 'observation statement') to (or near to) 1 – the effect of this modification is propagated over the rest of the agent's beliefs by conditionalization. My proposal is that observational acts – specifically, ones that induce agents to modify their probabilities directly[6] – be explicitly counted as one form of evidence within the Bayesian scheme (though I do not repudiate the existence of propositional evidence). (Clearly, in speaking of an observational act as a form of evidence, I mean to refer to the sensory content of the act. I will continue to use the term 'observational act' to refer to such sensory content below.) Insofar as the role of such acts within Bayesian epistemology has always been clearly essential, I am proposing nothing genuinely new. However, one gets a sense of how deeply entrenched the purely propositional conception of evidence is in the Bayesian community by the preliminary definition of evidence that motivates the old evidence problem – this is the claim that e is evidence for h iff $p(h/e) > p(h)$. The not so subtle implication is that the only thing worthy of the name 'evidence' is the sort of thing on which a probability function may be defined – a proposition.

To posit observational acts as a form of evidence is to take one small step forward in forging a solution to the quantitative problem: suppose e is a proposition that is evidence for h for A in virtue of A's belief in e, but suppose that A's justification for e consisted in A's having performed an observation O that establishes e to be true. A natural interpretation of what it means to assume that A's justified belief in e is 'deleted' is that O was not

---

[6] See Jeffreys (1965: 169) for an account of how observation causes probabilities to be 'directly' modified.

performed. Here there is no gross indeterminacy of the sort that befuddled the propositional conception of evidence: observational acts can be 'deleted' (by assuming they were not performed) in a determinate way. One might object that the assumption of the non-performance of some observational act O is just as indeterminate as the supposition that A does not believe e – for just as there are many sub-justification probabilities to choose from when imagining that A does not believe e, so there are many other acts of observation A may suppose would have been performed had O not been performed. This objection, however, misunderstands my proposal: in imagining that A imagines O not to have been performed I do not propose that A imagines that some other observation was performed in its place. Rather, I propose that we consider what probability function A now believes would be the appropriate one on the assumption that O had not been performed and no other observation had been performed in its place.

Where e is a belief which is evidence for h, then, and the epistemic basis for e consists of the past performance of some observational act(s) OBS, then the degree to which belief in e supports h can be computed by subtracting from the current probability of h the probability of h that would be rational had OBS not been performed. This amounts to a solution to the problem of the indeterminacy of p∗(e). But this proposal raises a hard question. Insofar as the assumption that OBS was not performed generates a determinate value for p∗(e), this must be because the agent possesses independent evidence pertinent to e other than OBS, evidence that is contained in the agent's background belief K. Thus the agent who has performed OBS actually possesses a complex mixture of evidence pertinent to e – the OBS evidence and the K-evidence (now imagining the K evidence to be purged of the influence of OBS). But why, exactly, do we regard the probability of e based on K (but not OBS) as the relevant probability for e on the proposal that 'e is not possessed'? What is the principled distinction between the OBS evidence and the K-evidence that establishes that the degree to which belief in e supports h is measured by the difference made to the probability of h by the OBS evidence (if indeed this is true)? Let us call this 'the distinction problem' – it is a problem unaddressed, I believe, by existing literature. We will confront it squarely below.

7.2 H-INDEPENDENT E-EVIDENCE: EINSTEIN IN 1915

Let us consider one particular agent: Einstein at a particular point during 1915. Einstein by this point has long since proposed his general theory of relativity (GTR) and in this year has already showed that GTR (plus relevant auxiliary information) entails M = the Mercury perihelion is advancing at 574 arcs/seconds per century. M had previously been established (by a complex series of observations and mathematical computations), and it provided an important piece of evidence for GTR. We assume that Einstein considers GTR less than fully proven at this point – there remains some probability that Newtonian Mechanics (NM) (the main contender to GTR) will prove ultimately true. There is a small probability, we assume he believed in 1915, that neither theory is true. Our question is: to what degree did the belief in M confirm GTR for Einstein at some point during 1915 after the derivation of M from GTR?

Much of the discussion about this example has focused on the problem of the indeterminacy of p∗(M) – as Glymour notes (1980: 88), the historical process by which M was established as true was long and complex, and it is far from clear how we should interpret the assumption that, e.g., Einstein did not believe M in 1915. Applying the evidence deletion approach, how much of the actual historical process should we assume did not occur? The apparently arbitrary nature of any answer to this question was one of the reasons Glymour declared the quantitative problem unsolvable. I think it would illuminate the present problem if we temporarily set this problem aside – with the promise to return to it later. To do so I propose that we adopt a strong simplifying assumption which is utterly counter to the historical facts – it is: (Q) The truth of M was established by a single, discrete observational act o-m prior to 1915 – and there was (for Einstein in 1915) no observational evidence for M independent of o-m. Assuming Q allows us to put aside one common form of criticism of attempts to solve the quantitative old evidence problem on the basis of the evidence deletion interpretation of the e-difference approach – and I think it is an interesting question whether the problem is independently solvable. Assuming Q, it would seem that the degree to which belief in M confirms h is equal to p(GTR) – p∗(GTR), where p(GTR) was Einstein's probability for GTR in 1915 based on his background belief K (which included M) and p∗(GTR) is the probability Einstein believed would be correct if K were unchanged except that it was adjusted so as to incorporate the assumption that o-m was not performed. In fixing the probability of p∗(GTR) we can scarcely avoid the question what value should be assigned to p∗(M) – but this would seem to be straightforwardly computable as follows:

$$(*) \ p^*(M) = p^*(M/GTR)p^*(GTR) + p^*(M/NM)p^*(NM)$$
$$+ p^*(M/\sim \& \ GTR \ \& \sim NM)[1 - p^*(GTR) - p^*(NM)]$$

Clearly p∗(M/GTR) is near one given the 1915 derivation of M from GTR and justified auxiliaries (though not one, as we assume that the relevant auxiliaries, though believed, have probability less than one). p∗(M/NM) was surely very low since there was precious little reason to think that M should be true if NM was true (though, since M is neither a contradiction nor is ruled strictly impossible by any background belief, this term will not receive probability 0). Now an additional reason Glymour regards p∗(M) as difficult to compute is that it is wholly unclear how the last term p∗(M/∼GTR & ∼NM) is to be computed – let us call this 'the problem of the catchall.' Glymour concedes that one might try to ignore this term but thinks there is no Bayesian motivation for doing so. I agree the term cannot be ignored but I see no serious threat here – I present my proposal for handling this problem in a footnote.[7]

Glymour's (1980) is the mother of all old evidence papers, and its skepticism about the solvability of the quantitative problem comes down to the view that (∗), although it is intuitively the sort of measure that characterizes the relevant non-justification probability for M, suffers from assorted technical problems that are probably collectively fatal (including the problem of the catchall and the indeterminacy of p∗(e)). But, setting aside these various technical problems as we have, the question now should be raised on what basis (∗) establishes the relevant non-belief probability for M. Consider again Einstein in 1915 after the M-derivation (and assuming (Q)): as matters then stood, Einstein was possessed of a complex body of evidence pertinent to M. He was possessed of the belief that the

---

[7] I propose that in the absence of any observational evidence for p∗(M/∼GTR & ∼NM) there is no basis for assigning any probability to this conditional probability – and thus would simply assign the unit interval [0,1] to p∗(M/∼GTR & ∼NM). The effect of this would be that p∗(e) would also take an interval value, ranging from the probability justified by the assignment of 0 to the last term to that justified by the assignment of 1. Insofar as our only purpose in identifying p∗(M) is to stipulate the condition under which M is not believed, it seems to me perfectly acceptable to concede that this condition does not involve the attribution of a point valued probability to M but an interval value. In this event, the value p∗(GTR) will likewise take an interval value and the effect of coming to believe M (once o-m is performed) will be that the interval probability p∗(GTR) is replaced with the point valued p(GTR) (since the interval p∗(M) will be replaced by the point valued probability p(M)). This would admittedly require a corresponding modification of the manner in which the degree of support provided by the belief in M to GTR is computed. I propose that we may think of 'degrees of support' as likewise taking on interval values. Where the endpoints of the interval valued probability for p∗(M) are x and y, the degree of support provided by belief in M to GTR will be the interval [x',y'] such that x' = p(GTR) – p∗(GTR/Pr(M)=x) and y' = p(GTR) – p∗(GTR/Pr(M)=y). I see no threatening problem or paradox arising from the proposal that degrees of support take on interval values. To the contrary, to the extent that the value of p∗(M) is indeterminate, the degree to which belief in M confirms GTR should be indeterminate to a corresponding degree. (There is considerable literature that develops the notion of interval valued degrees of belief – see the references cited by Howson and Urbach 1989: 69.)

observational act o-m had been performed, but also possessed of the
evidence necessary for the computation of p∗(M). This latter evidence
we might call the 'p∗(M)-evidence': it consists of all his relevant background belief except for the information conveyed by the act o-m. This will
include all relevant evidence pertinent to the truth of GTR, of NM, and the
auxiliary information necessary for the computation of the probabilities
p(M/GTR) and p(M/NM), assuming all such evidence has been purged of
any impact of the observational act o-m. Now in computing the degree to
which belief in M confirms GTR for Einstein at this point we proceed to
compare his current probability for GTR with the probability he would
assign to it if o-m had not been performed. We thus identify the state in
which Einstein does not possess M with the state in which he possesses the
p∗(M) evidence but not the o-m evidence. But at this point we encounter
the distinction problem: what is the motivation for identifying the state in
which Einstein does not believe M with the one in which he possesses the
p∗-evidence but not the o-m evidence? Why identify Einstein's non-belief
in M with this state, rather than one in which he has only some of the
p∗(M)-evidence and no o-m evidence (or some other state)? In what
follows I turn to this question.

My proposal is that, in measuring the contribution of the so-called 'belief
in e' to the probability of h, we seek to measure the epistemic impact of that
body of evidence which supports e independently of h. Evidence E supports
e independently of h iff the probability of e, given E and assuming ∼h, is
higher than the probability of e when E is not given and ∼h is assumed. E
supports e by way of h iff E supports e but does not support e independently of h. E supports e by way of h, thus, because E supports h and h
supports e (i.e. p(e/h) > p(e)). For example, assuming the Michelson-
Morley experiment ('MME,' which demonstrated the non-existence of
an ether wind) supported GTR, then MME supported M – since GTR
(plus auxiliaries) entails M. But MME supports M only by way of GTR:
assuming ∼GTR, MME does not support M at all (i.e., p(M/MME &
∼GTR) = p(M/∼GTR). On the other hand the observation o-m supports
M independently of GTR, since the probability of M is greater upon the
performance of o-m than if o-m were not performed even assuming
∼GTR. Thus the solution to the distinction problem, with respect to
some justified e which is intuitively evidence for some h, focuses on the
distinction between e-evidence that supports e independently of h and that
e-evidence which supports e by way of h. Intuitively, this makes great sense:
that e-evidence which supports e by way of h has, in an obvious sense,
nothing to do with the proposition e itself – such e-evidence merely

confirms h which in turn makes e more probable. By contrast, e-evidence that supports e independently of h is precisely the sort of evidence to regard as constituting evidence for h in virtue of our belief in e – it is evidence for h that is channeled 'through' the proposition e, so to speak.

If this solution to the distinction problem is correct, it follows of course that in claiming that we seek to measure the degree to which 'belief in e' confirms some hypothesis h, we are speaking loosely – we ought to claim that we seek to measure the degree to which h-independent e-evidence supports h. This is my view of what earlier authors on the quantitative problem of old evidence have implicitly been seeking.

Thus, to return to our primary example, when the observational act o-m is our only evidence for M that supports M independently of GTR, it makes sense to interpret 'the degree to which belief in M supports GTR' as 'the degree to which GTR is confirmed by the act o-m.' It makes equally good sense to ignore what we called the 'p∗(M)-evidence' insofar as this is evidence pertinent to GTR which does not bear on M in any direct (i.e. GTR independent) way. The value accorded to p∗(M) is thus a reflection of the evidence for (and against) GTR (which will coincide largely with evidence against (and for) NM, since these are the two competitors for Einstein in 1915). It is reached by purging Einstein's overall belief structure of any evidence that confirms M independently of GTR.

Needless to say, a correct solution to the distinction problem is only half the battle in the quest for objective values for p∗(e). For there remains the problem of how the h-support provided by the evidence that supports e independently of h is to be measured. In cases in which the h-independent e-evidence is an observational act(s) (as in the example discussed above) we can easily measure the evidential impact of this evidence on h by comparing the current probability of h with the probability we would assign to h if that act(s) had not been performed.

The problem, of course, is that evidence for e need not exist only in the form of observational acts – it may also exist in the form of other beliefs. Suppose that A's belief in e is grounded on A's belief in a set of propositions p1,. . ., pn – how are we to presume that A's justification in e is deleted? The only viable interpretation of hypothetical evidence deletion that I can imagine involves the hypothesis that certain (actually performed) observational acts were not performed. We could cogently delete A's evidence for e insofar as there is some set of observational acts, O(e), which constitute the h-independent evidence for e insofar as it is ultimately this set of acts which serve to justify A's belief in p1,. . . pn. Under what conditions will such a set of observational acts exist? It seems to me that the availability of such a set of acts is assured insofar as we assume a foundationalist conception of the structure of justification. According to this conception, A's justification for e will form a pyramid-like structure with e at the top – A's belief in e rests upon A's belief in other propositions, which may in turn rest upon other beliefs, until we reach a set of 'basic beliefs' which are justified non-inferentially – and which are typically thought of as justified 'directly by experience' in some sense. But of course, foundationalism is quite controversial as a theory about the structure of justification. Critics of foundationalism are fond of pointing out that the sensory content of some observational act does not by itself confirm any particular statement – it is only by the lights of background theory B (part of A's background belief K) that such confirmation is possible. This apparently undercuts the possibility of any foundation for justification – hence the appeal of a coherentist theory of justification which claims that individual beliefs are not tested one by one against experience. Rather, it is entire systems of beliefs which are tested against experience – individual beliefs are accepted or rejected on the basis of how well they cohere with other beliefs in the system. On such a view, there is no set of experiences which count as the ones confirming some particular proposition (cf., e.g., Bonjour 1976, 1978). But if the coherentist is right, it seems to me, the evidence deletion approach to the quantitative problem will be defunct, precisely because there will be no uniquely identifiable set of observations which 'confirm e.' Thus the viability of the evidence deletion approach, as developed here, can be only as plausible as a foundationalist account of the structure of justification.

Unfortunately, even assuming a foundationalist account of justification, there is no guarantee that it will always be possible to solve the quantitative old evidence problem by this procedure. The reason is that the assumption that none of the observational acts in $O(e)$ were performed may strip the agent of far more information than was intended. Her belief in various pieces of relevant information other than e may somehow rest on the acts in $O(e)$. To see this, let us return to the example of the confirmation of GTR by Einstein's belief in M in 1915, this time surrendering the simplifying assumption Q and considering the actual historical process by which M was established. The e-difference approach would answer that this degree is equal to $p(GTR) - p*(GTR)$. The question, of course, is what value should be assigned to $p*(M)$. Now if we were to apply the above technique in determining $p*(M)$, we would consider what probability Einstein would assign to M if Einstein were stripped of all GTR-independent evidence for M – which will ultimately amount to the assumption that a determinate set of observations constituting the observational action totality O(M) were not performed. Now there was no single event which determined M to be true. Rather, there was a complicated series of computations of the Mercury perihelion advance stretching from Leverrier's computation in the middle of the nineteenth century to the early twentieth century computations of Doolittle and Grossman – some of these conflicted with one another, others were in agreement (Glymour 1980: 88). In supposing that M was not observed to be true, we are apparently to assume that the various observations of Mercury's orbit over a long period of time (on which the computations of the advance were based) were not performed. But how are we to reconcile this with the assumption that the planet Mercury is (in the relevant situation in which M is not confirmed by observation) nonetheless believed to exist, and believed to have a determinate mass and a determinate position and velocity at some time (since all this information is needed to compute the essential conditional probabilities like p∗(M/GTR)?[8] If we allow that sufficiently many observations of the planet Mercury were performed to establish these facts, it is quite unclear that we have excluded all observations that confirm M, insofar as the allowed observations are consistent with M (assuming both M and the other observations are correct) but not with many other possible orbits. Let us call the attempt to stipulate precisely which observational acts are to be removed from the corpus of observational acts 'the non-observation problem.' The problem afflicting the evidence deletion approach seems, in a case like this one, to be rather grave.

I see no adequate solution to the non-observation problem in this case, though I do not want to foreclose the possibility of one. And the fact that the solution may run aground on this type of case should not obscure the fact that there are many cases in which the solution will not run aground – these will be those in which the relevant class of observational acts can be removed without deleting important evidence for beliefs other than e. But in any case the only result of its running aground in this case is to cast suspicion on the attempt to solve the quantitative problem in every case in terms of the e-difference approach. Nothing said so far prohibits a measurement of the contributions of GTR-independent M-evidence to the probability of GTR by some approach other than the e-difference

---

[8] For to surrender these beliefs would be to suppose that Einstein faced a very different picture of the solar system than he in fact faced. If there is no knowledge of the very existence of the planet Mercury, for example, then any observed phenomenon which resulted from Mercury's presence (on the sun or other planets) would be anomalous and call into question the adequacy of both NM and GTR in unintended ways.

approach. We turn now to develop another method for solving this problem.

According to another method, the current example would be handled as follows: Einstein in 1915 (after the derivation of M from GTR) endorses probability function p on the basis of his background belief K, which includes belief in M. To compute the degree to which the GTR-independent evidence for M supports GTR, Einstein would consider what function p' he would endorse on the basis of K', where K' is produced by modifying K only by rendering the Mercury perihelion evidentially irrelevant to GTR. How would Einstein formulate K'? It would suffice for him to reason as follows: let Einstein assume that his picture of the universe is exactly as it is at t (i.e. as represented by K) but for one change: he assumes that the shape and dimensions of the Mercury perihelion are (and have been) affected not only by the usual factors (e.g. the mass of Mercury, the gravitational force of the sun and other actual bodies) but may have been affected by forces deployed by an omnipotent God (forces of unknown magnitude and direction) applied to the planet Mercury but nowhere else. Einstein assumes that God acted so as to insure the truth of M. God's action, Einstein assumes, took one of two forms: either God saw that M would be true without any interference on Her part, in which case God did nothing at all, or God foresaw that without intervention the Mercury orbit would have some dimension other than what it has, in which case God deployed forces on the planet Mercury (but nowhere else) that induced the planet to take on its actual orbit. Thus God acted (or chose not to act) so as to assure the truth of M – I refer to such an act as one of 'divine M-assurance.' Given the assumption of divine M-assurance, the Mercury perihelion will be regarded by A as simply irrelevant to the probability of GTR given divine M-assurance which assures us that p'(M/GTR) = p'(M/~GTR)). However, all independent evidence for or against GTR possessed by A at t is completely unaffected by the assumption of God's possible action. On this admittedly imaginative basis A will choose function p'; the relevance of the Mercury perihelion to GTR for A at t will be equal to p(GTR) – p'(GTR).

In general, where h is a theory that purports to provide an explanation of some experimentally determined fact e, we may compute the degree to which existing h-independent e-evidence supports h by subtracting from the current probability of h, assuming e and no divine e-assurance, the probability of h assuming divine e-assurance. As above, divine e-assurance determines that e must be true no matter whether h is true or not – and proceeds either by God's non-action (when She foresees that existing conditions will suffice to make e true) or God's action (when God foresees that existing conditions will not suffice to make e true). We assume God intervenes as little as possible in order to make e true – thereby disturbing the overall activity of the universe as little as possible in establishing the truth of e.

Why is divine e-assurance effective as a measurement of the h-independent e-evidence's contribution to the probability of h? Ultimately, the epistemic impact of such evidence on the probability of h is constituted by the epistemic impact of a particular set of observational acts on h's probability. Now when we seek to measure the degree to which an observational act confirms a hypothesis, we have two options: (1) we compare the probability of h on the assumption the act was performed with the probability of h on the assumption the act was not performed, or (2) we can compare the probability of h on the assumption that the performed act was relevant to a particular degree to h with the probability of h on the assumption that the act is stripped of its relevance to h (though nonetheless performed). An observational act is stripped of its relevance to h when the informational content of the act is stripped of any probabilistic correlation to h. (1) is the way of the e-difference approach, (2) the way of divine e-assurance. Both strategies serve to measure the impact of a set of observations on some h by comparing the probability of h assuming those observations are taken into account with the probability of h assuming those observations are ignored. There are, then, at least two ways of ignoring an observational act: one may pretend it wasn't made, or pretend it carried no relevant information. In cases in which both methods are usable, they should deliver the same results.

The relative advantage of divine e-assurance is that it does not require any observation surgery – it allows the agent's body of observations to remain unperturbed. Furthermore, I argued above that the e-difference approach depends for its viability on a foundationalist picture of justification, and its success is thus tied to the success of that picture. Divine e-assurance carries no commitment to any account of justification – and so much the better for it.

Likewise, where the agent regards e as confirming h because e and h are what might be called 'effects of a common cause,' divine e-assurance may be used to good effect in measuring the degree to which h-independent e-evidence confirms h. For example, let h = John has a sore throat and e = John has spots on his face – A regards e as confirming h insofar as A regards e and h as effects of a common cause (the measles). Suppose A believes e (by an act of observation) and regards e as confirming h – in determining the degree to which the act (which we assume is the only h-independent e-evidence A possesses) confirms h, A may subtract from his current probability for h that probability he would assign to h if he assumed divine e-assurance and thus regarded the truth of e as irrelevant to the probability of h.

But there are cases in which this method is clearly inapplicable. Clearly, it will not be applicable in cases in which believed e confirms h for A because A regards e as a possible cause of h. For example, let e = John smokes a pack of cigarettes a day and h = John will die of lung cancer. Clearly, we cannot measure the degree to which A's belief in e confirms h for A at some time by comparing A's current probability for h with what probability A would assign to h assuming divine e-assurance – that God somehow acted to insure the truth of e does not by any means render e epistemically irrelevant to h, precisely because e confirms h for A because A regards e as a possible cause of h. Clearly, divine e-assurance is no panacea for the quantitative problem of old evidence in general, though it is none-theless of considerable usefulness. But in cases like these there will either be some fact of the matter about the objective propensity of h given e (and operative background assumptions), or there will not be. If there is such a fact, then the degree to which e confirms h will reflect the degree of causation involved (ranging from 0 (no causal efficacy) to 1 (strict causal determination)) – if there is no such fact, then there is no fact of the matter to what degree e confirms h, and thus it is no fault of any confirmation measure that it does not render one.

7.4 TWO SOLUTIONS TO THE QUANTITATIVE PROBLEM

Two interesting proposals for solving the quantitative problem of old evidence are presented in David Christensen (1999) and Eells and Fitelson (2000). I will summarize each of these proposals and then com-pare them to my own. Christensen notes that on the standard distance measure of confirmation (which measures the degree to which E supports H as $p(H/E) - p(H)$) the degree to which E confirms H is sensitive to two factors: the degree to which E is linked to H, and the distance E has to travel in moving from $p(E)$ to probability 1. But Christensen's intuition is that the latter factor is irrelevant with respect to the question about the degree to which E confirms H. Why, indeed, should the degree to which

E confirms H have anything to do with whatever our current probability for E happens to be? He thus proposes a normalized version of the distance measure that eliminates the influence of the distance factor from the assessed confirmation of H by E. His measure S(E,H) measures the support afforded by E to H as follows:

$$S(E, H) = [p(H/E) - p(H)]/p(\sim E).$$

Christensen proves that $S(E,H) = p(H/E) - p(H/\sim E)$ (providing $p(E)$ takes neither 0 or 1 as values). Thus where H entails E (and thus $p(H/\sim E) = 0$), $S(E,H) = p(H/E)$. Because it renders the support provided by E to H invariant under changes in $p(E)$, this offers a neat solution to the quantitative problem of old evidence, for $S(E,H)$ can be high (if $p(H/E)$ is high) even if $p(E)$ is very high (provided $p(E) \neq 1$).

Before proceeding, it should be noted that Christensen concedes that S suffers from a problem that he believes is fatal – I will refer to this as the problem of redundant evidence. His example runs as follows: let H be the hypothesis that there are deer in some woods, D reports the discovery of deer droppings and A reports the discovery of a shed antler in these woods. Individually, D and A provide equally good (and nearly conclusive) proof of H. Suppose an agent comes to observe the droppings and thus raises D to a probability near one – she thus accords H also a probability near one. As the agent walks along, however, she comes to observe a shed antler and thus raises the probability of A to near one. Intuitively D and A are equally good, albeit redundant, pieces of evidence for H. However, once D is known, the value of S(A,H) drops to near zero (since $p(H/A) - p(H/\sim A)$ is near zero). Thus S fails to capture our intuition about the degree to which A confirms H. As Christensen explains, S cannot distinguish in this context between D, a known piece of evidence, and the general body of background knowledge with respect to which conditional probabilities like $p(H/A)$ and $p(H/\sim A)$ are computed. He suspects that this failure will afflict other measures as well.

Eells and Fitelson (2000) (hereafter, E & F) respond to Christensen by proposing another solution to the quantitative problem of old evidence that avoids the problem of redundant evidence. The gist of their idea is that our intuition that A is strong evidence for H can be captured by using a probability function that is based on background knowledge that does not incorporate D – but their choice of the relevant function is a particular actual past function of the agent's (rather than a counterfactual one). Suppose that an agent has over time come to possess various pieces of evidence $E_1, \ldots E_n$ that are each relevant to hypothesis H, and assume that these pieces of evidence "came in" (E & F's phrase) at times $t_1, t_2, \ldots t_n$ respectively (so that now it is just after t_n). Let p_0 be the subjective probability assignment that the agent begins with just before t_1. We can ask what the evidential relevance of some E_i is for H relative to some subset of the propositions $E_1, \ldots E_{i-1}, E_{i+1}, \ldots E_n$. Let B range over the 2^{n-1} conjunctions of the members of these 2^{n-1} subsets. E&F then offer the following definition:

Definition: E_i is evidence for H, relative to B, if and only if $p_0(H/B \& E_i) > p_0(H/B)$; and the degree of E_i's evidential support for H, relative to B, is the difference between these two probabilities.

Thus – to return to Christensen's example – our intuition that the degree to which A (the shed antler) supports H (the hypothesis that there are deer in the woods) is high can be captured by this method – E & F's definition will show that A strongly supports H relative to some B that does not include D.

I will respond to E & F's solution first. There is something correct, I believe, about E & F's point that the question to what degree a piece of evidence confirms a hypothesis is often a question that implicitly refers to some contextually determined body of background belief – our intuition that A strongly supports H is based on our belief that A would raise the probability of H dramatically if there were no other conclusive evidence for H present. But we also have another intuition – that A does not strongly support H after all, given the belief in D. E & F have managed to capture both intuitions without appeal to any counterfactual probability function (in contrast to my own proposals) – they do this by reaching back in time to an agent's past function p_0. I do agree that it is best to avoid if possible the use of counterfactual claims in our analysis of fundamental concepts like evidence, given the notorious ambiguities of such claims. In this case, however, it seems to me that the attempt to measure the degree to which a piece of evidence *currently* supports a hypothesis for an agent by making use of his *past* probability function seems fraught with danger. For, setting aside the agent's beliefs in $E_1, \ldots E_n$, what assurance do we have that the general background beliefs that undergird the distribution of probabilities in the agent's present function are the same, or even similar to, those that undergirded his past function? Background beliefs that determine the probabilistic relevance of any or all of the E_i could have changed dramatically. Thus it seems to me that there is no good reason to measure the agent's current degrees of support by way of one of his past probability functions – but this refutes E & F's proposal.

My own proposed solution to the problem of redundant evidence in Christensen's deer example is to say, when an agent assigns near one probability to A, D and H, that A is strong evidence for H insofar as $p(A) - p*(A)$ is high, where $p*$ is the probability function A would have if the observations that support A and D had not been performed. In a way this resembles E & F's solution – for it acknowledges that the perceived support of A for H depends on the exclusion of D from background belief. Likewise, the intuition that A does not support H strongly given belief in D can be captured on my version of the e-difference approach by choosing $p*$ by subtracting the A-supporting observations but not the D-supporters. But my proposal differs in its being a counterfactual solution: $p*$ is the function the agent would adopt if, contrary to fact, certain observations were 'deleted' from his body of observations. This avoids the problem that beset E & F's approach because it is grounded on the agent's current background belief (minus the particular pieces of evidence in question).

Some of the problems that have been directed to versions of the e-difference approach could be posed for E & F's solution. For example, in choosing the relevant function p_o E & F require that we consider a point in the agent's history just before the various E_i "came in" – but this simple phrase masks possible complexities in the historical process that could make trouble for their proposal as well. Suppose that the evidence for some E_i does not come in all at once but trickles in slowly – to what point in the agent's history do we recede in picking p_o? There remains, for that matter, the distinction problem, which E & F do not consider. But it seems to me that they could successfully supplement their own solution with the same sort of analysis I deploy on behalf of my own solution to remedy such objections. If this is the tack they follow, however, their proposal will end up just as committed to a foundationalist account of justification as my e-difference approach, and is no less likely to run aground on the non-observation problem. At such point I fall back on my theistic solution, which seems to me superior to E & F's solution no less than to my own e-difference solution.

Let us now turn to Christensen's measure S. As noted, S runs aground on the problem of redundant evidence. But for the moment I want to call attention to a different problem with S. S purports to measure the degree to which E confirms H – I take this to mean that S measures the degree of support received by H from E insofar as E is fully known or believed (and thus receives a probability near 1). So, presumably, if E has a near 1 probability, and S(E,H) is high, then H is receiving a high level of support from E by the lights of S. However, there are straightforward counterexamples to this. Take E (a logical consequence of H) to be an observation statement, and assume that H comes to have a probability near 1 on some basis E' (wholly different from E). E itself remains completely unobserved. However, since p(H) is near 1, and since E is a logical consequence of H, p(E) is also near 1 at this point (I assume we have no basis for believing E other than the evidence E' that supports H). According to S, then, H is receiving a high level of support from E. But this strikes me as clearly false: E in this context is not providing any support to H, despite its strong 'link' to H and despite the fact that E is strongly believed. This is because our evidence for E is entirely by way of H: there is no H-independent E evidence. My e-difference solution delivers the correct result in this case: because there is no H-independent E-evidence for E, E provides no support for H (i.e. p(H) – p*(H) = 0, where p* is the function in which E has no directly supporting observations – hence p* and p are the same function!). Because the results of my theistic solution should coincide with those of my e-difference solution, the theistic solution also delivers the right result.

Christensen's solution goes wrong in this case, it seems to me, because he begins with the conviction that the distance the probability of E must travel from some subjustification problem to 1 is irrelevant to the degree of confirmation provided to H by E. I agree that there need be no connection between an agent's 'current' probability for E, whatever it is, and E's confirmation of H. But there is a connection, as I have argued, between a particular subjustification probability for E and E's confirmation of H – it is the value of p*(E). This reflects the probability of E that is based entirely on evidence that comes by way of H. The lower this probability is, the greater the support provided by E that comes from the agent's acquiring a particular kind of support for E – that provided by H-independent E-evidence. As I argue above, this is the evidence whose impact is relevant to the question of E's ability to confirm H.

Now let us suppose that E receives direct support from some set of observations OBS – OBS by itself (without the evidence that comes by way of H) would suffice to give E a probability near one, so both E and E' are known on the basis of direct observation. Now it seems to be the case that E does provide support for H, as there is H-independent E-evidence that supports H in the form of OBS. Christensen's S measure delivers this result (since I am considering a case in which p(H/∼E) = 0, S(E,H) will be high despite the impact of OBS). However, my e-difference solution apparently delivers a result of very low support, since the effect of the presence of evidence E' is to render p(H) very close to p*(H). The problem my solution apparently runs aground on here is essentially the problem of redundant evidence. Ironically, given that this is the problem that Christensen concedes is fatal to S, S escapes the problem in this particular example. But it is important to remember that there are conflicting intuitions available about cases of redundant evidence. If E is conceived as background knowledge, then E' loses its probative weight, whereas if E is not so conceived, E' has probative weight. I have suggested above how I would address the problem of redundant evidence, and this is by choosing p* in such a way as to reflect the impact of whichever body of observations support a piece of evidence that is relegated to background knowledge (just as E & F's solution does) – the problem is structurally the same as Christensen's deer example.

7.5 IMPLICATIONS FOR BAYESIANISM?

As noted at the beginning of this chapter, Glymour's 1980 essay 'Why I am not a Bayesian' explained that one of the reasons he rejected Bayesianism as a research project in confirmation theory was that Bayesianism was afflicted, he thought, with an unsolvable problem: the quantitative problem of old evidence. But we need not reach as far back as 1980 to find philosophers who point to the unsolvability of the quantitative problem of old evidence as a good reason to reject Bayesianism as a whole – Jarret Leplin argues in his (1997: 45) that the inability of Bayesians to give an adequate solution to this problem (together with other reasons) should lead us to reject the adequacy of Bayesian confirmation theory as an analysis of confirmation. I take the point of this chapter to be that such if anyone dumped 'Bayesianism' (whatever this terms means, exactly) on such a basis, it would have been dumped for the wrong reason.

It is, in my view, one of the great misconceptions about the problem of old evidence that it is only a problem for Bayesians – if one is not committed to Bayesianism, one is free, supposedly, to ignore the problem of old evidence. But why is this? Is it because non-Bayesians have no interest in establishing whether or not some heretofore accepted belief, or heretofore performed observational act, is evidence for some hypothesis, or no interest in determining the degree to which that belief or act supports that hypothesis? Presumably not – such questions are very foundational questions about the nature of evidence that anyone interested in 'how knowledge is acquired' should regard as perfectly central. But what then, are the non-Bayesian answers to these questions? To put the point another way, does one really have to be a Bayesian to say that a piece of evidence confirms a hypothesis if we are more confident in that hypothesis if we possess the evidence than if we don't, or that the degree of confirmation provided by that evidence to some hypothesis can be computed by comparing the probability of that hypothesis based on the possession of the evidence with the probability of the hypothesis justified when ignoring the evidence? Surely not. But this means that anyone committed to such commonplace claims will be faced with the sorts of problems the account in this chapter means to address. 'The Bayesian problem of old evidence,' like 'the problem of the inner cities,' is badly named: it is a problem for everyone.

7.6 CONCLUSION

This chapter began by noting that when $p(e)$ is near one, the difference between $p(h/e)$ and $p(h)$ is very small – thus if we measure the degree to which e confirms h by the distance between $p(h/e)$ and $p(h)$ this confirmation is necessarily very small. This suggests a kind of predictivism – for if we think of novel evidence at t as evidence that is not known at t, the preference for novelty is explained by the larger difference between $p(h/e)$ and $p(h)$ for unestablished e. That there is something wrong with this argument seems clear, because it entails a patently false conclusion: that antecedently known e cannot strongly support any hypothesis (consider, for example, Mayo's SAT score example (1.2.3)). At this point it is clear where this argument goes wrong. First of all, the implicated conception of novelty turns out to be the old temporal notion of novelty which Lakatos apparently endorsed in his (1970) and (1971), but this is a notion rejected long ago for excellent reasons. Secondly, in setting the degree to which e confirms h equal to $p(h/e) - p(h)$, the argument fails to note that the confirming power of e can condition not only $p(h/e)$ but $p(h)$ itself. As we have seen, an accurate assessment of quantitative support for an agent requires the introduction of a counterfactual function like p* or p' which compares the value of $p(h)$ with what probability h would have if e were 'deleted' – I have argued in this chapter that such functions can be identified and used to good effect. Thus predictivism will have to be defended in other ways.

All of this bears importantly on key claims defended in this book – for these claims pertained to the support provided for particular theories by particular pieces of evidence not only at the moment such evidence was taken into account but for subsequent moments as well. For example, the claims of Chapter 3 that various skeptical evaluators were induced by the success of Mendeleev's predictions to revise their estimations of the probabilities of Mendeleev's background beliefs entailed that the success of these predictions constituted some kind of important evidence for those background beliefs over a period of time beginning in Stage 2, which commenced with the early confirmations of his predictions. Not only did such confirmations boost the actual probabilities assigned to such background beliefs (and to PL) at some point in time, but they went on being evidence with a particular degree of probative weight after such times as well.[9]

As noted at the beginning of this chapter, the quantitative problem of old evidence can be seen to pose a serious threat to the thesis of predictivism: if those who argue that there is no fact of the matter how strongly any known piece of evidence confirms a hypothesis are right, then the thesis of predictivism cannot be sustained. Predictivism requires the existence of objective degrees of quantitative support, as it is a thesis about how quantitative support varies between predicted and accommodated evidence. However, as we have seen, there is every reason to take seriously the idea that known evidence, be it predicted or accommodated, confirms a particular hypothesis to a particular degree. Predictivism, thus, is neither vindicated nor undermined by the problem of old evidence.

---

[9] Though of course it is reasonable that the degree to which some particular E confirms some T may change over time, as new evidence or background knowledge becomes available.

CHAPTER 8

A paradox resolved

8.1 INTRODUCTION

This book began with the observation that predictivism is, prima facie, a paradoxical thesis. By this point any appearance of paradox has fully disappeared. In this chapter I offer a concise summary of the central theses of this book, and what I take to be its contributions to the philosophy of science. These theses fall into three main areas: confirmation theory, the realist/anti-realist debate, and the methodology of science. I deal with each in turn.

8.2 CONFIRMATION THEORY

One central claim has been that the phenomenon of predictivism is a real one in scientific theory evaluation and that it takes on a variety of forms. The virtuous/unvirtuous distinction intersects with the tempered/thin distinction to produce four forms of predictivism – and there is considerable reason to think that a typical scientific community will have members that instantiate more than one form (Chapter 3).

My term 'unvirtuous predictivism' refers to a phenomenon which resembles something well charted in the philosophical literature. This refers to the tendency to view the accommodator (but not the predictor) with a certain kind of suspicion, as though the accommodator were something like a Popperian pseudoscientist – one who endorses too highly theories that have been built to fit some salient body of evidence (or endorsed because they fit this evidence). Something resembling this view is well articulated by, e.g., Lipton (2004: Ch. 10). However, the term 'virtuous predictivism' refers to a thus far relatively unnoted phenomenon. This refers to the tendency to admire the predictor over the accommodator though both are presumed logically impeccable in their endorsements. Closely related to this thesis is a new account of what constitutes a novel confirmation, one I have deemed 'endorsement novelty.'

Certainly, other theories of predictivism might be interpreted in a way that would allow them to 'cover' the phenomenon of virtuous predictivism. But this is because, as I see it, these theories are couched in sufficiently vague language as to allow any number of interpretations. Maher's claim that successful prediction can serve as evidence that the predictor 'has a reliable method' of making predictions might plausibly refer to either virtuous or unvirtuous predictivism, or both. Kahn, Landsberg and Stockman's claim that successful prediction can serve as evidence that the predictor is 'talented' could be cashed out in terms of an absence of adhocery or a body of true background beliefs on the predictor's part, as might White's position that prediction testifies that the predictor is 'reliably aiming at the truth.' By neglecting the virtuous/unvirtuous distinction, such accounts run the risk of suggesting that all there is to predictivism is some version of the good old prohibition on ad hoc hypotheses. A theory that is explained at a higher level of precision is Sober and Hitchcock's (2004) theory – but this account is clearly an account of unvirtuous predictivism.

It was noted in (3.8.4.2) that Worrall's theory of predictivism resembles my account of virtuous predictivism to a degree. Worrall claims that prediction carries special weight because prediction provides unconditional confirmation for a theory – this is true, he argues, because predictions have a tendency to confirm the 'core idea' of a theory that accommodations often lack. This claim resembles at least somewhat my exposition of virtuous predictivism, which attributes to prediction a greater tendency to confirm the background beliefs of the predictor. The differences between Worrall's account and my own were emphasized – as were my reasons for rejecting Worrall's account – in (3.8.4.2). Nonetheless it seems to me that Worrall's account is – in spirit anyway – the closest thing in existing literature to what I have called virtuous predictivism.

Another central project has been to proclaim the essentially social nature of predictivism. 'Social predictivism' connotes two distinct claims: first, predictivism occurs in its most important form in the context of a pluralistic environment in which theory evaluators make important use of the judgments of other scientists (Chapters 2 and 3). The epistemic significance of other scientists' judgments can be given a highly rigorous formulation (Chapter 6), thus making sense of other scientists' judgments as a legitimate and important form of scientific evidence. Secondly, the epistemic significance of a successful prediction depends critically on the size of the predicting community, the number of successful predictors, and on by what method the predictor(s) was located (Chapter 5). This resonates with the central role of the high frequency assumption in the realist/anti-realist debate (Chapter 4).

Although I have emphasized the social dimension of confirmation throughout this book, as I have emphasized the importance of epistemic pluralism, it should be emphasized one final time that the phenomenon of predictivism can be countenanced in terms accessible to the purely individualistic evaluator. This is because the epistemic individualist evaluator can in some cases discern a link between prediction and a feature that carries epistemic significance – and this is all 'thin' predictivism requires. The feature could be the presence of additional evidence, or the special confirmation of background belief – depending on the form of predictivism at stake.

Finally, I have noted the logical connections between predictivism and the quantitative problem of old evidence, and offered original solutions to the latter problem (Chapter 7). I also argue for the superiority of these solutions – the theistic solution in particular – over various competing solutions. Contrary to a certain impression the old evidence problem makes, the old evidence problem supplies no legitimate motivation for predictivism. It was also noted that the widespread conviction that the quantitative problem of old evidence is unsolvable – because there is allegedly no fact of the matter to what degree known evidence supports a hypothesis – appears to threaten predictivism (since predictivism posits facts about degrees of quantitative support). But there is indeed such a fact of the matter – and thus predictivism faces no such threat.

8.3 THE REALIST/ANTI-REALIST DEBATE

In Chapter 5, I note there to be two entirely distinct versions of the so-called miracle argument for scientific realism – the 'miraculous theory argument' and the 'miraculous endorsement argument.' There is a mountainous philosophical literature which purports to develop a version of the miraculous theory argument which emphasizes the need to explain the novel successes of science, but this argument is logically fallacious. Closely related to this fallacious argument is an argument for strong predictivism, which is fallacious essentially for the same reason. While philosophical literature abounds with passionate rejections of strong predictivism, much of this literature rejects this thesis for the wrong reason. The wrong reason is that strong predictivism entails biographicalism which is widely presumed to be a false thesis about theory confirmation. I of course have defended biographicalism at much length. The actual reason strong predictivism should be abandoned is that the only plausible argument for it involves a fallacious application of Ockham's razor.

The miraculous endorsement argument – whose principle proponent to date has been Boyd – is logically impeccable, though for it to succeed one must assume that empirically successful theories emerge at a rate too high to be the result of chance, a claim I have referred to as 'the high frequency assumption.' I have no justification for this assumption to offer. I have argued, however, that the realist can explain the high frequency assumption, if it is true, while there is no such explanation forthcoming from the anti-realist. Should the high frequency assumption turn out true, the realist should regard herself as greatly vindicated. But the more important results for our purposes of these considerations are these: (1) once the miraculous theory argument is rejected in favor of the miraculous endorsement argument, an account of predictivism emerges which is essentially identical to the account developed in Chapters 2 and 3: successful predictions are, unlike accommodations, evidence for the truth of the predictor's background beliefs. And (2): predictivism, in both its virtuous and unvirtuous forms, is essentially available only to the realist. The anti-realist must be an anti-predictivist. This is because there is no principled way to argue that novel success is special evidence for empirical adequacy without also arguing that novel success is special evidence for truth.

Finally, Magnus and Callender's (2004) claim that the realist/anti-realist debate is (at least in its 'wholesale version') hopelessly entangled in the base rate fallacy is shown to be false. Epistemic pluralism offers hope for the reasonable prior probabilities that avoiding the base rate fallacy requires. Their attempt to vindicate 'retail realism' is shown vulnerable to various problems that can be overcome by the resources on offer here, particularly by the realist's solution to the problem of take-off theories. There is no cause for ennui here.

8.4 THE METHODOLOGY OF SCIENCE

There is a long standing tradition in the philosophy of science of epistemic individualism. This individualism claims that when an individual evaluates a theory's probability in a completely rational way, he or she does so only by relying on his or her own assessment of the relevant evidence and logic. Scientists, on this view, do not rely on the judgments of other agents in assessing theories insofar as they are being genuinely rational.

One consequence of epistemic individualism is that non-experts who lack the requisite knowledge cannot engage in rational theory evaluation.

On the face of it, this consequence might seem perfectly reasonable. Theory evaluation should be left to those expert enough to do it correctly. But on reflection this is a disturbing and unacceptable conclusion. If non-experts have literally no way to assess theories endorsed by the scientific community, this can only mean they have no reason to assess the activity of science itself as authoritative. This leaves them with little reason to support science or to abide by its recommendations. Rational arguments in favor of the theories recommended by the scientific community must be available to non-experts if respect for science among the general populace is to be sustained. Novel confirmation, among other things, can play an important role in non-expert evaluation – though other phenomena such as consensus are no less important. It seems to me that non-expert theory evaluation is fundamental to the success of science, and that its nature should fall firmly within the purview of the philosophy of science (Chapter 2).

Though many would agree that non-experts need a well developed epistemic pluralism to guide them, many would deny that experts themselves engage in pluralistic theory assessments. For it is surely the role and duty of the expert to 'think for herself' about the theories falling within her own domain of expertise. I have argued that this kind of thoroughgoing expert individualism is sheer dogma – experts themselves must make use of the judgments of other experts even within their own fields. This is because many experts are, as I have deemed them, team members or interdisciplinary experts, humble experts, imperfect experts, and reflective experts. All of these have no recourse but to look to the judgments of other experts for critical epistemic information. The methodology of science is far more pluralistic than is generally recognized (Chapter 2). This does not mean, of course, that experts defer entirely to other experts in forming such judgments – their own deliberations must be combined with those of others to form aggregated judgments that make use of all relevant information (Chapter 6). It is in this pluralism, once more, that the solution to the paradox of predictivism is primarily found.

The appeal of epistemic individualism lies partly in the cloud of suspicion that appeals to authority have been under throughout the history of Western science. A well known illustration of this is the story of Galileo, whose published defense of the Copernican worldview led him to be tried by the Inquisition and ultimately to a sentence of life imprisonment. Stillman Drake explains that

The age into which Galileo was born was one in which the power of authority was uppermost in every sphere of activity – political, religious, and philosophical. It was therefore virtually impossible to attack that power in one sphere without disturbing it in others. To Galileo it was clear that in matters of scientific investigation, authority as such could not be allowed any weight; observation, experiment, and reason alone could establish physical truth. (1967: 64), quoted in Walton (1997: 46–47)

This appearance of antithesis between genuine science and the appeal to authority survives to this day in the philosophy of science (despite welcome exceptions, particularly among feminist philosophers of science such as Longino (1990)). But once it is conceded that one may appeal to authorities who are themselves possessed of substantial knowledge of the results of observation, experiment, and reason, this antithesis should of course dissolve – along with the paradox of predictivism.

Glossary

Agnostic evaluator – An evaluator who does not know whether the endorsers she faces are virtuous or unvirtuous.

Anti-realist challenge – The claim that anti-realism can explain the truth of the high frequency assumption by attributing empirical adequacy – but not truth – to the background beliefs of the scientific community. This challenges the miraculous endorsement argument for realism, which argues that only the imputation of truth to such background belief can explain the high frequency assumption.

Anti-superfluity principle (ASP) – A version of Ockham's razor which maintains that putative explanations should not be endorsed when they are explanatorily superfluous vis à vis the truths they purport to explain, or when they contain explanatorily superfluous components.

Biographicalism – The thesis that biographical facts about persons carry epistemic significance in theory evaluation.

Comprehensive – A belief set is comprehensive with respect to some theory T if that set is rich in informative content that bears on the probability of T.

Distinction problem – The problem of determining why observational evidence for e, but not theoretical evidence, should be deleted in considering what probability e should be assigned if it were not known.

Divine E-assurance – The assumption that God acted (or chose not to act) so as to insure the truth of E.

E-difference approach – The attempt to determine the degree to which E supports T by comparing the probability of T on the assumption that E is known with the probability of T on the assumption that E is not known.

Epistemic individualism – (1) the thesis that rational agents do not assign epistemic significance to the judgments of other agents in assessing theory probability but base such evaluation only on the relevant evidential information, or (2) the procedure of not deferring to such judgments but considering only the relevant evidential information on one's own in theory evaluation.

Epistemic pluralism – (1) the thesis that rational agents do assign epistemic significance to the judgments of other agents in assessing theory probability, or (2) the procedure of assigning epistemic significance to such judgments.

Endorsement novelty – The claim that N (a consequence of T that is established to be true) counts as a novel confirmation of T relative to agent X iff X posts

246

an endorsement-level probability for T (i.e. a probability greater than or equal to L) that is based on a body of evidence that does not include observation-based knowledge of N.

Endorser – One who posts an endorsement level probability for a theory (i.e. posts a probability at or above L (cf. 2.2)).

Evaluator – One who assesses the probability of a theory based on some body of evidence – which may include whether the evidence was predicted or accommodated by some endorser of that theory.

Genetic theory of predictivism – The thesis that predictivism is true because successful prediction, unlike accommodation, constitutes evidence that the theory was generated by a dependable source.

High frequency assumption – The claim that empirically (especially novelly) successful theories emerge from the scientific community at a rate too high to be the result of chance.

Humble expert – An expert who formulates a judgment about the probability of a theory about which she is less than wholly certain, due to the possibility of error on her part (due, e.g., to forgetfulness or computational fallibility).

Imperfect expert – An expert who lacks perfect knowledge of his own field of scientific specialization.

Indeterminacy of $p^*(e)$ – the apparent indeterminacy regarding which probability should be assigned to some known e on the assumption that e is not known.

Keynesian dissolution of the paradox of predictivism – The claim that predictivism is an illusion caused by a failure to note the effect of preliminary evidence that supports a theory that makes a prediction.

Miraculous endorsement argument for realism – The argument which claims that realism is the only position that does not make the fact that scientists have been successful in endorsing theories that subsequently proved successful miraculous (since realism imputes reliability to the procedures of scientific theory appraisal).

Miraculous theory argument for realism – The argument which claims that realism is the only position that does not make the empirical successes of particular theories miraculous (since realism imputes approximate truth to such theories).

Non-observation problem – The problem of stipulating which (actual) observational acts should be supposed not to have occurred on the assumption that some statement e has not been observed to be true.

Prediction per se – the fact that an endorser endorsed a theory without appeal to N (a consequence of the theory), and thus predicted N. A prediction per se can carry epistemic significance when it indicates the presence of reasons for endorsement possessed by the endorser (e.g., in the form of the endorser's background beliefs).

Predictive success – the fact that a prediction made by an endorser is established to be true. Predictive success can serve to confirm the truth of the background beliefs that prompted the endorser to make the prediction.

Problem of take-off theories – The problem of providing a miracle-free account of the emergence of take-off theories (viz., the 'first true theories' that emerged at a take-off point in the history of science (cf. 4.3.2)).

Reflective expert – An expert in some field who seeks to know what reason there is for him to believe in his own background beliefs as they bear on that field.

Sufficient – A belief set is sufficient with respect to some theory T iff the beliefs in the set are both true and comprehensive enough to accord an informative probability for T.

Strong predictivism – The thesis that prediction, in and of itself, carries epistemic weight. Strong predictivism entails biographicalism.

Tempered predictivism – The form of weak predictivism which asserts that facts about prediction and accommodation have epistemic significance for theory evaluators in virtue of their symptomatic roles, i.e., in virtue of their serving as indicators of some other feature of theories that carries epistemic weight.

Theory emergence – The point in the history of science at which a theory is assessed by at least some competent scientists as sufficiently plausible to be worth further investigation – it thus receives a certain preliminary form of endorsement from the scientific community.

Thin predictivism – The form of weak predictivism which denies that facts about prediction and accommodation have any epistemic significance even in virtue of their symptomatic roles, i.e., in virtue of their serving as indicators of some other feature of theories that carries epistemic weight. Thin predictivism asserts only that there is a link between prediction and some such feature.

Unvirtuous agent – an agent who posts a probability (say, for a theory) that is not actually justified based on the evidence available to her and her background belief.

Unvirtuous predictivism – the thesis that predictivism is true because accommodators are (unlike predictors) prone to post probabilities for the theories they endorse that are too high, given their actual evidence and background belief.

Use-novelty – The claim that N (a consequence of T that has been established to be true) is a novel confirmation of T iff T was not built to fit N, i.e., N was not 'used' in constructing T.

Virtuous agent – an agent who posts a probability (say, for a theory) that is actually justified on the basis of the evidence possessed by that agent and her background beliefs.

Virtuous predictivism – the thesis that predictivism is true because predicted evidence (1) indicates the existence of background beliefs which prompted the prediction and (2) has a stronger tendency to confirm the endorser's background beliefs than accommodated evidence.

Weak predictivism – The thesis that prediction carries no epistemic weight in and of itself but is linked to some other feature of theories that does carry epistemic weight. Weak predictivism comes in two forms: tempered predictivism and thin predictivism.

Bibliography

Achinstein, P. (1991), *Particles and Waves: Historical Essays in the Philosophy of Science*, New York and Oxford: Oxford University Press.

— (2001), *The Book of Evidence*, Oxford University Press.

Alston, W. (1976), "Two Types of Foundationalism," *Journal of Philosophy* 73: 165–168.

Bamford, G. (1993), "Popper's Explications of Ad Hocness: Circularity, Empirical Content, and Scientific Practice," *British Journal for the Philosophy of Science*, 44: 335–355.

Barnes, E. (1996a), "Discussion: Thoughts on Maher's Predictivism," *Philosophy of Science*, 63: 401–410.

— (1996b), "Social Predictivism," *Erkenntnis*, 45: 69–89.

— (1998), "Probabilities and Epistemic Pluralism," *British Journal for the Philosophy of Science*, 48: 31–47.

— (1999), "The Quantitative Problem of Old Evidence," *British Journal for the Philosophy of Science*, 48: 31–47.

— (2000), "Ockham's Razor and the Anti-Superfluity Principle," *Erkenntnis*, 53: 353–374.

— (2002), "Neither Truth nor Empirical Adequacy Explain Novel Success," *Australasian Journal of Philosophy*, 80 (4): 418–431.

— (2002), "The Miraculous Choice Argument for Realism," *Philosophical Studies*, 111 (2): 97–120.

— (2005a), "Predictivism for Pluralists," *British Journal for the Philosophy of Science*, 56: 421–450.

— (2005b), "On Mendeleev's Predictions: Comment on Scerri and Worrall," *Studies in the History and Philosophy of Science*, 36: 801–812.

Barret, J. (1996), "Oracles, Aesthetics, and Bayesian Consensus," *Philosophy of Science*, Supplement to 63, S273–S280.

Bensaude-Vincent, B. (1986), "Mendeleev's Periodic System of Chemical Elements," *British Journal for the History of Science*, 19: 3–17.

— (2000), "From Teaching to Writing: Lecture Notes and Textbooks at the French Ecole Polytechnique," in Lundren and Bensaude-Vincent, pp. 273–294.

Bonjour, L. [1976], "The Coherence Theory of Empirical Knowledge," *Philosophical Studies*, 30: 281–312.

Bonjour, L. [1978], "Can Empirical Knowledge Have a Foundation?," *American Philosophical Quarterly*, 15: 1–13.

Boyd, R. (1981), "Scientific Realism and Naturalistic Epistemology," *PSA 1980* Vol. 2, East Lansing: Philosophy of Science Association, pp. 612–662.

    (1985), "*Lex Orandi est Lex Credendi*," in Churchland and Hooker (eds.), pp. 612–662.

    (1991), "The Current Status of Scientific Realism," in Boyd, R., Gasper, P., and Trout, J. D., (eds.) *The Philosophy of Science*, Cambridge: MIT Press, pp. 195–222. Also printed in Leplin, J. (ed.) 1984, *Scientific Realism*, Berkeley: University of California Press, pp. 41–82.

Brush, S. (1996), "The Reception of Mendeleev's periodic law in America and Britain," *Isis*, 87: 595–628.

Bynum, W. F., Browne, E. J., and Porter, R. (1981), *Dictionary of the History of Science*, Princeton: Princeton University Press.

Campbell, R. and Vinci, T. (1983), "Novel Confirmation," *The British Journal for the Philosophy of Science*, 34: 315–341.

Cassebaum, H. and Kauffman, G. B. (1971), "The Periodic System of the Chemical Elements: The Search for its Discoverer," *Isis*, 62: 314–327.

Chihara, C. [1987], "Some Problems for Bayesian Confirmation Theory," *The British Journal for the Philosophy of Science*, 38: 551–560.

Churchland, P. and Hooker, C. (eds.) (1985), *Images of Science: Essays on Realism and Empiricism with a Reply from Bas C. Van Fraassen*, Chicago and London: The University of Chicago Press.

Christensen, D. (1999), "Measuring Confirmation," *Journal of Philosophy*, 96: 437–461.

Clark, F. W. (1884), *Elements of Chemistry*, New York: Appleton.

Collins, R. (1994), "Against the Epistemic Value of Prediction Over Accommodation," *Nous*, 28: 210–224.

Coady, C. A. J. (1992), *Testimony: A Philosophical Study*, Oxford: Clarendon Press.

Darden, L. (1991), *Theory Change in Science: Strategies from Mendelian Genetics*, New York and Oxford: Oxford University Press.

de Milt, C. (1951), "The Conference at Karlsruhe," *Journal of Chemical Education*, 28: 421–424.

Douven, I. (2000), "The Anti-Realist Argument for Underdetermination," *The Philosophical Quarterly*, 50: 371–375.

Drake, S. (1967), "Galileo Galilei," *The Encyclopedia of Philosophy*, Vol. 3, P. Edwards (ed.), New York: Macmillan.

Earman, J. (ed.) (1983), *Testing Scientific Theories*, Minneapolis: University of Minnesota Press.

    (1992), *Bayes or Bust? A Critical Examination of Bayesian Confirmation Theory*, Cambridge, MA: MIT Press.

    (1993), "Carnap, Kuhn and the Philosophy of Science Methodology," in P. Horwich (ed.), *World Changes*, Cambridge, MA, MIT Press.

Earman, J. and Glymour, C. (1980), "Relativity and Eclipses: The British Eclipse Expeditions of 1919 and Their Predecessors," *Historical Studies in the Physical Sciences*, 11: 49–85.

Edwards, P. (1967), *The Encyclopedia of Philosophy*, New York: Macmillan Publishing Company and the Free Press.

Eells, E. (1990), "Bayesian Problems of Old Evidence," in Savage, W. (ed.), *Scientific Theories*, Minneapolis: University of Minnesota Press.

Eells, E. and Fitelson, B. (2000), "Measuring Confirmation and Evidence," *Journal of Philosophy*, 97: 663–672.

Farrar, W. V. (1965), "Nineteenth Century Speculations on the Complexity of the Chemical Elements," *The British Journal for the History of Science*, 2: 291–323.

Fine, A. (1986), "Unnatural Attitudes: Realist and Instrumentalist Attachments to Science," *Mind*, 95: 145–179.

　(1991), "Piecemeal Realism," *Philosophical Studies*, 61: 79–96.

Fitelson, B. (1999), "The Plurality of Bayesian Measures of Confirmation and the Problem of Measure Sensitivity," *Philosophy of Science*, Supplement to Vol. 66, No. 3, S362–S378.

Foley, R. (2001), *Intellectual Trust in Oneself and Others*, Cambridge: Cambridge University Press.

Frankel, H. (1979), "The Career of Continental Drift Theory," *Studies in History and Philosophy of Science*, 10: 21–66.

Garber, D. (1983), "Old Evidence and Logical Omniscience in Bayesian Confirmation Theory," in Earman [1983].

Gardner, M. (1982), "Predicting Novel Facts," *British Journal for the Philosophy of Science*, 33: 1–15.

Giere, R. (1983), "Testing Theoretical Hypotheses," in J. Earman (ed.) *Testing Scientific Theories*, Minnesota Studies in the Philosophy of Science, Vol. 10, Minneapolis: University of Minnesota Press.

　(1984), *Understanding Scientific Reasoning*, 2nd edn., New York: Holt, Rinehart and Winston.

Gillespie, C. C. (ed.) (1971), *Dictionary of Scientific Biography*, New York: Charles Scribner's Sons.

Glymour, C. (1980), *Theory and Evidence*, Princeton: Princeton University Press.

Good, I. J. (1967), "The White Shoe is a Red Herring," *British Journal for the Philosophy of Science*, 17: 322.

Gutting, G. (1985), "Scientific Realism versus Constructive Empiricism," in Churchland and Hooker (eds.) pp. 118–131.

Grünbaum, A. (1984), *The Foundations of Psychoanalysis: A Philosophical Critique*, University of California Press.

Hacking, I. (1979), "Imre Lakatos's Philosophy of Science," *British Journal for the Philosophy of Science*, 30: 381–410.

Hardwig, J. (1985), "Epistemic Independence," *Journal of Philosophy*, 82: 335–349.

　(1991), "The Role of Trust in Knowledge," *Journal of Philosophy*, 88: 693–708.

Harker, D. (2006), "Accommodation and Prediction: The Case of the Persistent Head," *The British Journal for the Philosophy of Science*, 57: 309–321.

Hobbes, J. (1993), "Ex Post Facto Explanations," *Journal of Philosophy*, 93: 117–136.

Hoel, P. G. (1971), *Introduction to Mathematical Statistics*, New York: John Wiley and Sons.

Holden, N. E. (1984), "Mendeleev and the Periodic Classification of the Elements," *Chemical International*, 6: 18–31.

Howson, C. (1984), "Bayesianism and Support by Novel Facts," *British Journal for the Philosophy of Science*, 35: 245–251.

(1985), "Some Recent Objections to the Bayesian Theory of Support," *The British Journal for the Philosophy of Science*, 36: 305–309.

(1988), "Accommodation, Prediction, and Bayesian Confirmation Theory," *PSA 1988*, 2: 381–292.

(1990), "Fitting Your Theory to the Facts: Probably Not Such a Bad Thing After All," in W. Savage (ed.), *Scientific Theories*, Minnesota Studies in the Philosophy of Science, vol. 14. Minneapolis: University of Minnesota Press.

(1991), "The 'Old Evidence' Problem," *The British Journal for the Philosophy of Science*, 42: 547–555.

(1996), "Error Probabilities in Error," *Philosophy of Science*, Supplement to Vol. 64, pp. S185–S194.

Howson, C. and Franklin, A. (1991), "Maher, Mendeleev and Bayesianism," *Philosophy of Science*, 58: 574–585.

Howson, C. and Urbach, P. (1989), *Scientific Reasoning: The Bayesian Approach*, Open Court, La Salle.

Ihde, A. J. (1961), "The Karlsruhe Congress: A Centennial Retrospective," *Journal of Chemical Education*, 38 (2): 83–86.

Jeffrey, R. (1965), *The Logic of Decision*, McGraw-Hill Book Company.

Kahn, J. A., S. E. Landsberg, and A. C. Stockman (1990), "On Novel Confirmation," *British Journal for the Philosophy of Science*, 43: 503–516.

Kantorovich, A. (1993), *Scientific Discovery: Logic and Tinkering*, State University of New York Press.

Kaplan, M. (1996), *Decision Theory as Philosophy*, Cambridge: Cambridge University Press.

Keynes, J. M. (1929), *A Treatise on Probability*, London: Macmillan and Co.

Kitcher, P. (1989), "Explanatory Unification and the Causal Structure of the World," in Kitcher, P. and Salmon, W. (eds.), *Scientific Explanation*, Minneapolis: University of Minnesota.

(1993), *The Advancement of Science: Science Without Legend, Objectivity Without Illusions*, New York and Oxford: Oxford University Press.

Knight, D. M. (1970), *Classical Scientific Papers – Chemistry*, 2nd Series, New York: American Elsevier.

(1992), *Ideas in Chemistry: A History of the Science*, New Brunswick, NJ: Rutgers University Press.

Kuhn, T. S. (1970), *The Structure of Scientific Revolutions*, 2nd edn., Chicago: University of Chicago Press.

Kukla, A. (1998), *Studies in Scientific Realism*, New York and Oxford: Oxford University Press.

Kultgen, J. H. (1958), "Philosophic Conceptions in Mendeleev's Principles of Chemistry," *Philosophy of Science*, 25: 177–183.

Kusch, M. (2002), *Knowledge by Agreement*, Oxford: Clarendon Press.

Lackey, J. and Sosa, E. (eds.) (2006), *The Epistemology of Testimony*, Oxford: Oxford University Press.

Ladyman, J., Douven, I., Horsten, L., and van Fraassen, B. (1997), "A Defense of Van Fraassen's Critique of Abductive Inference: Reply to Psillos," *The Philosophical Quarterly*, 47: 305–321.

Laudan, L. (1981), *Science and Hypothesis*, Dordrecht: D. Reidel.

(1984), "Explaining the Success of Science: Beyond Epistemic Realism and Relativism," in Cushing, J., Delaney, C. F., and Gutting, G. (eds.) *Science and Reality*, Notre Dame: University of Notre Dame Press.

(1990), "Normative Naturalism," *Philosophy of Science*, 57: 44–59.

Lakatos, I. (1970), "Falsification and the Methodology of Scientific Research Programmes," in I. Lakatos and A. Musgrave (eds.), *Criticism and the Growth of Knowledge*, Cambridge: Cambridge University Press, pp. 91–196.

(1971), "History of Science and its Rational Reconstructions," *Boston Studies in the Philosophy of Science*, 8: 91–135.

Laing, M. (1995), "The Karlsruhe Conference, 1860," *Education in Chemistry*, 32: 151–153.

Lange, M. (2001), "The Apparent Superiority of Prediction to Accommodation as a Side Effect: a Reply to Maher," *British Journal for the Philosophy of Science*, 52: 575–588.

Lehrer, K. (1974), "Induction, Consensus and Catastrophe," in Bodgan, R. J. (ed.) *Local Induction*, Dordrecht: D. Reidel.

(1976), "When Rational Disagreement is Impossible," *Nous*, 10: 327–332.

(1977), "Social Information," *Monist*, 60: 473–487.

(1978), "Consensus and Comparison: A Theory of Social Rationality," in Hooker, C. A., Leach, J. J., and McClennan, E. F. (eds.), *Foundations and Applications of Decision Theory*, Vol. 1, Dordrecht: D. Reidel.

Lehrer, K. and Wagner, C. (1981), *Rational Consensus in Science and Society*, Dordrecht: D. Reidel.

Leplin, J. (1975), "The Concept of an Ad Hoc Hypothesis," *Studies in the History and Philosophy of Science*, 5: 309–345.

(1982), "The Assessment of Auxiliary Hypotheses," *British Journal for the Philosophy of Science*, 33: 235–249.

(1987), "The Bearing of Discovery on Justification," *Canadian Journal of Philosophy*, 17: 805–814.

(1997), *A Novel Defense of Scientific Realism*, New York and Oxford: Oxford University Press.

(2000), "The Epistemic Status of Auxiliary Hypotheses," *The Philosophical Quarterly*, 50: 376–379.

Lewis, D. (1979), "Counterfactual Dependence and Time's Arrow," *Nous*, 13: 455–476, reprinted in Lewis, D. (1986) *Philosophical Papers*, Vol. II, Oxford University Press, pp. 32–52.

Lipton, P. (1991), *Inference to the Best Explanation*, London: Routledge.

(1992–3), "Is the Best Good Enough?," *Proceedings of the Aristotelian Society*, 93: 89–104.

(2004), *Inference to the Best Explanation*, 2nd edn., London and New York: Routledge.

Locke, J. (1995), *An Essay Concerning Human Understanding*, Amherst, NY: Prometheus Press.

Longino, H. (1990), *Science as Social Knowledge*, Princeton: Princeton University Press.

Lundgren, A. (2000), "Theory and Practice in Swedish Chemical Textbooks during the Nineteenth Century," in Lundgren and Bensaude-Vincent, pp. 91–118.

Lundgren, A. and Bensaude-Vincent, B. (2000), *Communicating Chemistry: Textbooks and Their Audiences 1789–1939*, USA, Science History Publications.

MacDonald, H. (1985), *Eating for the Health of It*, Priddis, Alberta: Austin Books.

Magnus, P. D. and Callender, C. (2004), "Realist Ennui and the Base Rate Fallacy," *Philosophy of Science*, 71: 320–338.

Maher, P. (1988), "Prediction, Accommodation, and the Logic of Discovery," *PSA 1988*, 1: 273–285.

(1990), "How Prediction Enhances Confirmation," in J. M. Dunn and A. Gupta (eds.), *Truth or Consequences: Essays in Honor of Nuel Belnap*, Dordrecht: Kluwer, pp. 327–343.

(1993), "Howson and Franklin on Prediction," *Philosophy of Science*, 60: 329–340.

[1996], "Subjective and Objective Confirmation," *Philosophy of Science*, 63: 149–174.

Manahan, S. E. (1997), *Environmental Science and Technology*, Boca Raton, New York: Lewis Publishers.

Marsden, G. (1980), *Fundamentalism and American Culture: The Shaping of Twentieth Century Evangelicalism*, New York and Oxford: Oxford University Press.

Maxwell, G. (1962), "The Ontological Status of Theoretical Entities," in H. Feigl and G. Maxwell (eds.), *Minnesota Studies in the Philosophy of Science*, Vol 3. Minnesota: University of Minnesota Press, pp. 12–13.

Mayo, D. (1991), "Novel Evidence and Severe Tests," *Philosophy of Science*, 58: 523–553.

(1996), *Error and the Growth of Experimental Knowledge*, University of Chicago Press.

McAllister, J. W. (1996), *Beauty and Revolution in Science*, Ithaca and London: Cornell University Press.

Meehl, P. (1999), "How to Weight Scientists' Probabilities is Not a Big Problem: Comment on Barnes," *British Journal for the Philosophy of Science*, 50: 283–295.

Mendeleev, D. (1889), "The Periodic Law of the Chemical Elements: Faraday Lecture," printed in Knight (1970), pp. 322–344.

Merton, R. (1973), "The Normative Structure of Science," printed in Merton, R. *The Sociology of Science: Theoretical and Empirical Investigations*, ed. N. W. Storer, Chicago: University of Chicago Press.

Mill, J. S. (1961), *System of Logic, Ratiocinative and Inductive*, 8th edn., London.

Murphy, N. (1989), "Another Look at Novel Facts," *Studies in the History and Philosophy of Science*, 20. (3), 385–388.

Musgrave, A.. (1974), "Logical versus Historical Theories of Confirmation," *British Journal for the Philosophy of Science*, 25: 1–23.

(1985), "Realism versus Constructive Empiricism," in Churchland and Hooker (eds.), pp. 197–221.

(1988), "The Ultimate Argument for Scientific Realism," in R. Nola (ed.), *Relativism and Realism in Science*, Dordrecht: Kluwer Academic Publishers, pp. 229–252.

Nickles, T. (1987), "Lakatosian Heuristics and Epistemic Support," *British Journal for the Philosophy of Science*, 38: 181–205.

(1988), "Truth or Consequences? Generative Versus Consequential Justification in Science," *PSA 1988, Vol. 2*, 393–405.

Niiniluoto, I. (1983), "Novel Facts and Bayesianism," *British Journal for the Philosophy of Science*, 34: 375–379.

Nunan, R. (1984), "Novel Facts, Bayesian Rationality, and the History of Continental Drift," *Studies in the History and Philosophy of Science*, 15: 267–307.

Partington, J. R. (1964), *A History of Chemistry*, Vol. 4, London: Macmillan and Co. LTD.

Peirce, C. S. (1897), Anonymous Review of Mendeleev's Principles of Chemistry, *Nation*, 25: 424.

Peters, D. and Ceci, S. (1982), "Peer Review Practices of Psychological Journals: The Fate of Published Articles Submitted Again," *Behavior and Brain Sciences*, 5: 187–195.

Popper, K. (1963), *Conjectures and Refutations: The Growth of Scientific Knowledge*, New York and Evanston: Harper & Row.

Putnam, H. (1975), *Mathematics, Matter, and Method*, New York: Cambridge University Press.

Psillos, S. (1996), "On Van Fraassen's Critique of Abductive Reasoning," *The Philosophical Quarterly*, 46: 31–47.

(1997), "How Not to Defend Constructive Empiricism: A Rejoinder," *The Philosophical Quarterly*, 47: 369–372.

(1999), *Scientific Realism: How Science Tracks Truth*, London and New York: Routledge.

Quine, W. V. O. (1951), "Two Dogmas of Empiricism," *The Philosophical Review*, 60: 20–43. Reprinted in Quine (1953), *From a Logical Point of View*, Harvard University Press.

(1960), *Word and Object*, Cambridge: MIT Press.

Redhead, M. (1986), "Novelty and Confirmation," *British Journal for the Philosophy of Science*, 37: 115–118.

Richmond, A. (1999), "Between Abduction and the Deep Blue Sea," *The Philosophical Quarterly*, 49: 86–91.

Rosenkranz, R. (1983), "Why Glymour IS a Bayesian," in Earman, 1983, pp. 69–98.

Salmon, W. (1966), *The Foundation of Scientific Inference*, University of Pittsburgh Press.

(1981), "Rational Prediction," *British Journal for the Philosophy of Science*, 32: 115–125.

Sambursky, S. (1971), "Structure and Periodicity: Centenary of Mendeleev's Discovery," *Proceedings of the Israel Academy of Sciences and Humanities*, 4: 1–13.

Scerri, E. (2005), "Response to Barnes' critique of Scerri and Worrall," *Studies in the History and Philosophy of Science*, 36: 813–816.

(2007), *The Periodic Table: Its Story and its Significance*, Oxford and New York: Oxford University Press.

Scerri, E. and Worrall, J. (2001), "Prediction and the Periodic Table," *Studies in the History and Philosophy of Science*, 32: 407–452.

Schaffner, K. (1974), "Einstein versus Lorentz: Research Programmes and the Logic of Theory Evaluation," *British Journal for the Philosophy of Science*, 25: 45–78.

Schlesinger, G. (1987), "Accommodation and Prediction," *Australasian Journal of Philosophy*, 65: 33–42.

Shapere, D. (1982), "The Concept of Observation in Science and Philosophy," *Philosophy of Science*, 49: 485–525.

Shapin, S. (1994), *A Social History of Truth*, Chicago and London: University of Chicago Press.

Shapin, S. and Schaffer, S. (1985), *Leviathan and the Air-Pump: Hobbes, Boyle and the Experimental Life*, Princeton University Press.

Sobel, M. (1987), *Light*, Chicago: University of Chicago Press.

Sober, E. (1988), *Reconstructing the Past: Parsimony, Evolution and Inference*, Cambridge, MA, and London: MIT Press.

Sober, E. and Hitchcock, C. (2004), "Prediction Versus Accommodation and the Risk of Overfitting," *British Journal for the Philosophy of Science*, 55: 1–34.

Solov'ev, Y. I. (1984), "D.I. Mendeleev and the English Chemists," *Journal of Chemical Education*, 61: 1069–1071.

Soyfer, V. (1994), *Lysenko and the Tragedy of Soviet Science*, New Brunswick, NJ: Rutgers University Press.

Speelman, C. (1998), "Implicit Expertise: Do We Expect Too Much From Our Experts?," in Kirsner, K., Speelman, C., Maybery, M., O'Brien-Malone, A., Anderson, M., MacLeod, C. (eds.), *Implicit and Explicit Mental Processes*, Mahwah, NJ, London: Lawrence Erlbaum Associates, Publishers, pp. 135–148.

Spring, R. J. (1975), "Vindicating the periodic law," *Education in Chemistry*, 12: 134–138.

Stanford, P. K. (2000), "An Anti-Realist Explanation of the Success of Science," *Philosophy of Science*, 67: 266–84.

Strathern, P. (2000), *The Quest for the Elements*, New York: St. Martin's Press.

Suppe, F. (1998), "The Structure of a Scientific Paper," *Philosophy of Science*, 65 (3): 381–405.

Thomason, N. (1992), "Could Lakatos, Even With Zahar's Criterion for Novel Fact, Evaluate the Copernican Research Programme?," *The British Journal for the Philosophy of Science*, 43: 161–200.

Van Fraassen, B. (1980), *The Scientific Image*, Oxford: Clarendon Press.

(1988), "The Problem of Old Evidence," in D. F. Austin (ed.)., *Philosophical Analysis*, Dordrecht: Kluwer Academic.

(1989), *Laws and Symmetry*, Oxford: Clarendon Press.

Velikovsky, I. (1950), *Worlds in Collision*, New York: Doubleday.

(1955), *Earth in Upheaval*, New York: Doubleday.

Venable, F. P. (1896), *The Development of the Periodic Law*, Chemical Publishing Co., Easton, PA.

van Spronson, J. W. (1969), *The Periodic System of Chemical Elements: A History of the First Hundred Years*, Amsterdam, London and New York: Elsevier.

Wagner, C. (1978), "Consensus Through Respect: A Model of Rational Group Decision Making," *Philosophical Studies*, 34: 335–349.

(1981), "The Formal Foundations of Lehrer's Theory of Consensus," in Bogdan, R. U. (ed.) *Profile: Keith Lehrer*, Dordrecht: D. Reidel.

(1997), "Old evidence and new explanation," *Philosophy of Science*, 64: 677–690.

(1999), "Old evidence and new explanation II," *Philosophy of Science*, 66: 283–288.

(2001), "Old evidence and new explanation III," *Philosophy of Science*, 68: s165–s175.

Walton, D. (1997), *Appeal to Expert Opinion: Arguments from Authority*, University Park, Pennsylvania: The Pennsylvania State University Press.

White, R. (2003), "The Epistemic Advantage of Prediction Over Accommodation," *Mind*, 112: 448, 653–683.

Worrall, J. (1985), "Scientific Discovery and Theory Confirmation," in J. C. Pitt (ed.), *Change and Progress in Modern Science*, pp. 301–331, D. Reidel Publishing Company.

(1989), "Fresnel, Poisson and the White Spot: The Role of Successful Predictions in the Acceptance of Scientific Theories," printed in Gooding, D., Pinch, T. and Schaffer, S. (eds.), *The Uses of Experiment: Studies in the Natural Sciences*, Cambridge University Press.

(2002), "New Evidence for Old," in P. Gardenfors, J. Wolenski, and K. Kihania-Placek (eds.), *In the Scope of Logic, Methodology and Philosophy of Science*, vol. 1, pp. 191–209, The Netherlands: Kluwer Academic Publishers.

(2005), "Prediction and the 'Periodic Law': A Rejoinder to Barnes," *Studies in the History and Philosophy of Science*, 36: 817–826.

Zahar, E. (1973), "Why did Einstein's Programme supercede Lorentz's?," *British Journal for the Philosophy of Science*, 24: 95–123.

(1989), *Einstein's Revolution*, La Salle, IL: Open Court.

Index

www.ingramcontent.com/pod-product-compliance
Ingram Content Group UK Ltd.
Pitfield, Milton Keynes, MK11 3LW, UK
UKHW020436180125

453697UK00006B/73